QUANTUM THEORY
AND BEYOND

QUANTUM THEORY AND BEYOND

ESSAYS AND DISCUSSIONS ARISING FROM A COLLOQUIUM

EDITED BY
TED BASTIN

CAMBRIDGE
AT THE UNIVERSITY PRESS
1971

Published by the Syndics of the Cambridge University Press
Bentley House, 200 Euston Road, London N.W.1
American Branch: 32 East 57th Street, New York, N.Y.10022

© Cambridge University Press 1971

Library of Congress Catalogue Card Number: 77-127237

ISBN: 0 521 07956 X

The sketch of Niels Bohr on
the frontispiece was drawn from
memory by O. R. Frisch

Printed in U.S.A.

CONTENTS

CONTENTS

LIST OF PARTICIPANTS

Y. AHARONOV, Yeshiva University, Amsterdam Avenue, Queens, New York City, U.S.A.

R. H. ATKIN, Department of Mathematics, University of Essex, Colchester, Essex.

E. W. BASTIN, Cambridge Language Research Unit, 20 Millington Road, Cambridge, England.

D. BOHM, Physics Department, Birkbeck College, Malet Street, London, W.C. 1.

J. BUB, Department of Physical Chemistry, University of Minnesota, Minneapolis, Minnesota, U.S.A.

M. BUNGE, Physics Department, McGill University, Montreal 2, Province of Quebec, Canada.

G. F. CHEW, Lawrence Radiation Laboratory, University of California at Berkeley, Berkeley, California, U.S.A.

O. R. FRISCH, Cavendish Laboratory, Free School Lane, Cambridge, England.

M. A. GARSTENS, Department of Physics, University of Maryland, College Park, Maryland, U.S.A.

H. J. GROENEWOLD, Institute for Theoretical Physics, Paddepoel, P.O. box 800, Groningen, Netherlands.

B. J. HILEY, Physics Department, Birkbeck College, Malet Street, London, W.C. 1.

C. W. KILMISTER, Department of Mathematics, King's College, Strand, London, W.C. 2.

D. S. LINNEY, Cambridge Language Research Unit, 20 Millington Road, Cambridge, England.

H. H. PATTEE, W. W. Hansen (Laboratories of Physics), Stanford University, Stanford, California 94305, U.S.A.

R. PENROSE, Department of Mathematics, Birkbeck College, Malet Street, London, W.C. 1.

A. PETERSEN, Yeshiva University, Amsterdam Avenue, Queens, New York City, U.S.A.

H. POST, Department of Philosophy, Chelsea College of Science and Technology, Manresa Road, London, S.W. 3.

PARTICIPANTS

G. M. PROSPERI, Universita Degli Scienze Fisiche, Viz Celoria 16, Milano, Italy.

J. ROTHSTEIN, Department of Computer and Information Science, Caldwell Laboratory, Ohio State University, Columbus, Ohio 43210, U.S.A.

C. F. FRHR. VON WEIZSÄCKER, Max-Planck-Institūt zur Erforschung der Lebensbedingungen der wissenschaftlich-technischen Welt, Munich.

J. H. M. WHITEMAN, University of Cape Town, South Africa.

PREFACE

The essays that form this book are based on papers read at an informal colloquium held at Cambridge in July 1968. The book takes the title of the colloquium which set before the participants the possibility that a paradigm change in quantum theory may be imminent. The colloquium was intended to provide opportunity, first to discuss the seriousness of the difficulties in quantum theory from points of view which do not presuppose the inevitability of the current approaches, and second, to discuss some possible alternative theories to see what a real change might involve.

Indications of the trends taken by the discussions have been given in short introductions to the different sections of the book. This division into sections approximately follows the order of proceedings of the colloquium.

Invitations to the colloquium were limited to twenty-five people, so that the participants could in fact discuss informally. This meant naturally that the selection had to be arbitrary, but an effort was made to see that all schools of thought which have seriously been put forward were presented.

The organizers of the colloquium were E. W. Bastin and D. Bohm. The general chairman was O. R. Frisch.

Grants toward the costs of the colloquium were made by The Royal Society, The Carnegie Institution of Washington, and Theoria Inc.— a small private trust.

I

INTRODUCTION

THE FUNCTION OF THE COLLOQUIUM
EDITORIAL

The foundations of quantum theory are different from any other part of physics in that they are in continual unrest. There is a problem of foundational quantum theory in a way that there is no problem about fluid dynamics, nor about crystallography nor even, though this is a more complex case, at the base of general relativity.† The task confronting this colloquium was really to locate this unrest unambiguously, or failing that, at least to have the various arguments that locate the source of the trouble in different places presented alongside one another.

There is a main-stream point of view on foundational quantum theory which is a crystallized form of the play of ideas current in the late twenties of this century and which forms the background to the teaching of physics. You might say that it is what the average physicist who never actually asks himself what he believes on foundational questions, is able to work with in solving his detailed problems. The terms of reference of the 'Quantum Theory and Beyond' colloquium were to examine this 'main-stream' view from standpoints independent of it, and the following comments give the reasons why it was thought this examination was desirable.

The main-stream view is not satisfactory to the logician, because it draws elements from several incompatible sources which the common sense and experience of the working physicist alone succeed in combining into an effective way of working. First among those sources is the detailed success of the schematization which quantum theory had produced by 1928 of the low energy spectra and therefore of atomic energy levels. With this schematization there was a method of calculating which worked and whose success could never be gone back on.

---

† There may be disagreement in general relativity on such a fundamental question, for example, as to how—if at all—to incorporate Mach's principle. Nevertheless once a decision to work in terms of a given set of postulates has been made, there is no problem in knowing what deductions do, and what do not, constitute valid parts of the theory.

1

However a difficulty now arose because of a fundamental change in outlook which the quantum theory made necessary. The new schemata required for their use that one cease to demand answers to questions which classical physics had regarded as essential to its very way of thinking that it should be able to answer. On the other hand no way of presenting the new schemata—not even the radically phenomenological method due to Heisenberg of operating with what was known about a quantum structure in matrix form—could stand on its own without drawing on thinking that was an integral part of the classical picture.

A *prima facie* circularity thus needs to be recognized in the dependence of quantum theory upon classical concepts. Quantum theory took its origins in an attempt to remedy a fundamental flaw in the classical physics (a flaw which could no longer be ignored once the 'ultra-violet catastrophe' in the theory of the spectral energy distribution had become fully apparent). Through its success in this original situation quantum theory came to be thought capable of subsuming classical theory as a special case, and capable, too, of applying its new and more profound concepts to make comprehensible a class of phenomena which were necessarily incomprehensible in the older picture. However, this expectation on behalf of the quantum theory is combined with the requirement that every quantum concept be operationally specified using classical arrangements of apparatus (and therefore—surely—of classical concepts) and therein lies the circularity.

To eliminate this circularity—dimly felt at first but issuing more clearly at the present time in flat contradictions—is the main problem of the foundations of the quantum theory. I call it 'the quantum problem'.

The next element in the main-stream view of quantum theory (and the one which has contributed most to it) was due to Bohr. Bohr developed a generalized philosophy of physical theory-making which has come to be called *complementarity*. His conception of the complementary relation of the wave and particle pictures, in—say—the two slit experiment with either photons or particles, is the best known example of the general idea, and certainly influenced its early history. Complementarity, in this specialized case, has precisely the effect of negating the classical demand for a progressively more refined specification of every object. Complementarity, by contrast, requires that the quantum object be described in each of two 'complementary' ways. To reach the 'wave–particle duality' as it is known to the experimentalist, however, much more is needed than this bald statement. It was necessary for Bohr to maintain the indivisibility of the whole, classically formulated, experimental arrangement that was needed to exhibit either a wave phenomenon or a particle phenomenon. Further, it was necessary to postulate that two such complete experimental arrangements could be mutually incompatible at a given time, in order to get the non-classical results.

The general principles: 'indivisibility of a total experimental arrangement', and 'mutual incompatibility of two such arrangements' have not penetrated the main-stream view, even though Bohr had found such an intellectual framework essential to the understanding of that complementarity. A much more widespread view is that quantum theory (particularly in the context of the Heisenberg uncertainty relation) exhibits an irreducible interdependence of that which is observed on the one hand and the corresponding act of observation on the other, and that the wave–particle duality has been shown by Bohr to be a consequence of this interdependence, though to worry just how he showed this would be overscrupulous. Of course this attitude is reasonable only if Bohr's background position is in fact compatible with current practice, and the assumption that would be made by the general body of physicists that their practice embodies the complementarity philosophy is questionable to say the least.

Efforts have been made to demonstrate the consistency of the quantum set of concepts with macroscopic concepts, given the Heisenberg uncertainty relation (consider for example the very thorough investigation of this type made by Rosenfeld and Bohr for the extension of quantization to fields). However, efforts of this kind are not now what is in question. The question actually is whether the Heisenberg relation itself is enough to add onto the physicist's conceptual armoury to turn a classical way of thinking into what Bohr would have thought adequate to that total nexus of apparatus and bits of the physical world that replaces the classical particle.

Wigner, Jauch and Yanase[1] put the matter of the intrusion of the observer in physical theory at the quantum level in a sharp form. 'There seems no escape from the...epistemological dilemma as long as one assumes the validity of the basic principles of quantum mechanics, in particular of the superposition principle and of the linearity of the equations of motion' (p. 5).

The 'epistemological dilemma' which these authors refer to arises in the first place out of the attempt to describe the process of measurement, where 'As has been pointed out many times before, von Neumann's theory, if followed to its ultimate consequences, leads to an epistemological dilemma'. They present two alternatives (p. 2):

The first alternative is to assume that quantum mechanics refrains from making any definite statement about 'physical reality' just as relativity theory has taught us to refrain from defining 'absolute rest'. Rather quantum mechanics only furnishes us with correlations between subsequent observations and these correlations are, in general, only statistical.

The second alternative is to say that the wave function, or more generally the state vector, is a description of the physical reality. In that case one has to admit that the state vector may change in two fundamentally different ways: continuously and causally, as a result of the lapse of time, according to the time-dependent Schrödinger equation, and discontinuously and erratically, as a result of observations. This second change is often called 'collapse of the wave-function'.

They conclude: 'According to the accepted principles of scientific epistemology, neither of these two points of view seems satisfactory.' For our present argument their conclusion is important in showing that there is an epistemological element—whether dilemma or not—within the complex of ideas that make up the main-stream view. There is also not much doubt that this epistemological element is what underlies the 'Quantum problem'.

The rest of the components of the main-stream view originated in attempts to say that the epistemological dilemma—properly viewed—is not a dilemma at all. The most direct way to do this is to show that the measurement process can be incorporated in a formal account of quantum theory like that of von Neumann. The various developments of quantum theory made with this aim are called 'measurement theory'.

Every point of view on the interpretation of the quantum formalism includes reference to probability and to statistical concepts. The wave-functions are universally thought to give statistical information about the movement of quantum particles, and what is known as the Born interpretation of the formalism consists in giving a primary interpretation to the product of conjugate complex wave-functions $\Psi\Psi^*$ for example as the square of the probability of finding a particle or photon at the point to which the wave-function refers. The Born interpretation of the wave-function came to be regarded by some as a complete statement of the epistemological situation in quantum theory, so that no further problems remained which might require such essentially non-classical concepts as the wave–particle duality for their comprehension.†

It is natural that people should have tried to say that the Born interpretation solves the quantum problem. The argument against this position is that in fact quantum particles have to be dealt with individually; a statistical treatment of them is adequate for some problems, but not for others.‡

Faced with the Uncertainty Relation, Einstein and Popper used Born's idea. Born had shown that quantum uncertainty arises—as does all uncertainty—from ordinary incompleteness of information: we only have statistical counts. But Heisenberg and Bohr were clear (and Born understood this) that the incompleteness of the information itself had a different origin from anything that one had hitherto met in statistical problems. We get an uncertainty in the presence even of maximal information in the quantum mechanical sense, and it is this possibility which is a quite new departure, for in pre-quantum statistical theory the existence of uncertainty could always be taken to show that information was not maximal.

The wish to regard the quantum ensemble as the interpretative unit for quantum theory—whether or not that position can be rigorously maintained—has had a considerable effect on the presentations of the theory, particularly by writers who are critical of the Copenhagen position. For example Blokhintsev, in a Marxist-philosophic article,[3] remarks that 'the Copenhagen school relegates to secondary importance the *statistical ensemble* and concentrates on the mutual relations of a single phenomenon and the instrument. This is an essential methodological error, for quantum mechanics becomes "instrumental" and the objective aspect of things is extinguished'.

To treat the information available at each stage of a quantum situation as a primary parameter is another approach which arises naturally in connexion with the quantum ensemble. Brillouin,[4] for

---

† Certainly not by Born himself however, for whom the statistical interpretation is not an alternative to complementarity but a further development of the same theory. He remarks [2], 'The true philosophical import of the statistical interpretation consists in the recognition that the wave picture and the corpuscle picture are not mutually exclusive, but are two complementary ways of considering the same process—a process whose accessibility to intuitive apprehension is never complete but always subject to certain limitations given by the principle of uncertainty'.

‡ Consider for example, that even if we do not accept the whole range of high energy phenomena as relevant in demonstrating concrete individuality of quantum systems, any interference experiment can be conducted at such low intensity that effects due to statistical assemblages of particles can be discounted, yet interference occurs just the same. Hence even in the traditionally central case of electron interference, purely statistical treatment is not possible.

example, treats the uncertainty principle—a physical limitation on observation—as a limitation on the availability of information. His underlying argument is that information is a numerical quantity and therefore in its nature susceptible to having limits imposed on it and that quantum theory tells us what the limits are. This argument does not answer the criticism that in the contexts to which the information concept naturally applies, limits to numerical quantities of information (as in communication channels) are man-made, whereas in the quantum situation their essential characteristic is their unalterableness.

The information concept appears to have one source other than the probabilistic aspect of the interpretation of the quantum state—namely the concept of *preparation* of the quantum state. In classical physics the history of a particle is essentially irrelevant to the understanding of the behaviour of that particle because we can, in principle, always go further in finding out more about the particle by procedures which can be carried out independently of the use we happen to be making of the particle. In quantum theory, this is not so, and some writers, for example Temple,[5] have developed the concept of the quantum state as a specification of what the particle—say—described by the state is known to have done. Thus if it has passed through three successive collinear slits in the presence of a magnetic field, then it is uncharged. (Perhaps this particular experimental procedure has selected *that* particle from a host of other particles with different endowments). Indeed when we specify the attributes of the particle—like its having no charge—then we are given shorthand specifications of just such pieces of knowledge. Hence the quantum state always refers back to earlier states, and can be regarded as a specification of the information that we have about the particle. This way of looking at things seems to give a powerful insight, but the link between information specified in this way, and information theory seems to need a great deal of further work if it is to prove really fruitful.

An interesting view which combines the idea of the preparation of a state with that of an ensemble, has been propounded by Fock,[6] who defines the state of a system in terms of both an initial and a final configuration, the latter being the ensemble of things which might have happened given what was known to start with.

Fock remarks: 'L'ensemble des possibilités potentielles pour l'expérience finale, qui découle de l'expérience initiale donnée, caractérise l'état d'un système. Si cette caractéristique est maximale (aussi complète que possible) on peut la nommer état du système tout court. L'expérience initiale peut être nommée maximale si elle conduit à la connaissance de l'état du système.'

To complete the list of foundational components of main-stream quantum theory, there is 'no-nonsense formalism': the view that a well defined mathematical scheme which works has been found and that is all we know and all we need to know. Some of the contributors to this book scrutinize the assumption which 'no nonsense formalism' makes, namely that the application of the quantum-theoretical rules is always unambiguous, so that it is always clear to the practitioner how he is to continue his analysis of every problem. For the moment, though, we will accept that everyone knows how to work with quantum mechanics. The von Neumann presentation comes nearest to reasoned backing for 'no-nonsense formalism'. However it is in character for the non-nonsense formalist to ignore the general opinion that for consistent thinking to be possible von Neumann's scheme needs to be completed by a satisfactory theory of measurement† or by a non-classical logic‡ or possibly by both (and for reasons that are certainly deep and will be discussed at length in papers and discussions in this book, neither of these is to hand).

Dirac, in a celebrated presentation of quantum theory,[9] gives a formalist account in which the difficulties over measurement and logic are settled in one stroke at the outset. The mathematical principle of superposition of states (which characterizes quantum theory as opposed to classical mechanics) is ascribed directly to the necessity of regarding a single particle or a single photon as being in each of two states prior to observation and as being definitely in one or other state as the instantaneous result of the observation. 'Observation' is not discussed by Dirac: he postulates no collapse of the wave-function which in other accounts is invoked to give some physical meaning to the alleged effects of the observation process. Probably Dirac's method,[10] reminiscent of one theological tradition for dealing with a mystery, is the best that can be done for no-nonsense formalism. He leaves the mystery, as stark as possible and says that the facts, in the physical case (it is revelation in the theological) demand it.

† Though measurement theory was not taken seriously by the founders of quantum theory, and Heisenberg[7] even says '...it would be a mistake to believe that this application of the quantum-theoretical laws to the measuring device could help to avoid the fundamental paradox of quantum theory'.

‡ It should be remembered that von Neumann himself[8] abandoned distributivity in the lattice of propositions that describe quantum states.

However one feels bound to protest that there is something very odd about renouncing the hope of understanding something and yet claiming to know that it is that something and nothing else to which experiment leads us. To put the point more sharply, no amount of experimental evidence can count in favour of a logical muddle.

Oddly, accounts like that of Dirac which put the intrusion of an epistemological element into the experimental situation most uncompromisingly even if most indigestibly† have been used by writers of a realist persuasion—particularly by Popper—to cut the formalism free from the regrettable epistemological elements so that quantum theory can be identified with the former and the latter ignored. Positions of this kind have had some influence on some physicists and therefore some influence on the 'main-stream view' of quantum theory. This influence has however probably been small because one can scarcely deny that the quantum theoretical formalism took the form it did precisely to accommodate those puzzling aspects of the quantal state of matter which this realist position seeks to find an excuse to ignore.

We now have the main strands that make up main-stream quantum theory. Each has its thorough-going adherents who want to achieve logical consistency at all costs in the terms of the strand of their particular choice and see all the other strands as convenient partial expressions or aspects of the correct picture whose validity extends just so far as each can be presented in the terms of their chosen strand. No such single strand has achieved a picture of the quantum situation which satisfies an overwhelming majority of physicists (and the 'Quantum Theory and Beyond' colloquium did not alter this negative state of affairs). One is therefore forced to contemplate a situation which must be unique in the history of science where the practitioners of a scientific theory which has reached the stage of being regarded as a finished product habitually work with a jumble of elements taken from a variety of different conceptual frameworks none of which, singly, is adequate to present the facts that are known, and each of which is partly or even largely incom-

---

† Dirac remarks, when discussing the experiment with a split beam of photons the components of which are brought to interference: 'Let us consider now what happens when we determine the energy in each of the components. The result of such a determination must be either the whole photon or nothing at all. Thus the photon must change suddenly from being partly in one beam and partly in another to being entirely in one of the beams. This sudden change is due to the disturbance in the translational state of the photon which the observation necessarily makes.'[10]

patible with the rest.† What keeps these practitioners feeling that it is one discipline they are operating is quite a puzzle. Partly it is a faith in the existence of a growing body of knowledge which constitutes physics and which persists massively unchanged whatever we may add onto it in the way of revolutionary principles or discoveries. Partly again it is the persistence of classical methods and concepts which have a degree of success which it is difficult to account for if the changes demanded by quantum theory are taken literally.

Either way, an unprecedented situation may be held to call for unprecedented treatment. Since Kuhn's *Structure of Scientific Revolutions* it has become commonplace to imagine scientific change as being typically large and abrupt rather than by steady progression. However for Kuhn a revolution or paradigm change can take place only when a new paradigm is clearly in the field. The unprecedented situation in quantum theory foundations may be held to be so extreme as to require the search for new paradigms, as it were from cold. Some of the contributors to 'Quantum Theory and Beyond' do hold just this, and have been prepared to stick their necks out in putting forward models which depart radically from quantum theory as it is usually known.

A little more needs to be said in explanation of this neck-sticking-out. Not long ago such efforts were met with complete incredulity on the part of physicists who were competent to evaluate such attempts. Current philosophy of science presupposed that distinct but comparable theories on fundamental issues could exist for empirical adjudication, and that such adjudication was the scientific norm, while in physical practice the edifice of physics was utterly monolithic, any significant conceptual change being always too radical in its effects for the imagination to be able to cope with the intermediate stage of conceptual confusion that it would produce.

Recently there has been widespread recognition that basic changes in physics may be necessary, and the incredulity has changed to scepticism, or perhaps to a feeling of powerlessness. The aim of 'Quantum Theory and Beyond' will have been completely achieved if the scepticism could be changed again to an atmosphere of reasoned assessment of the difficulties and obscurities that have to be overcome before any change in the basic structure of physical theory can

† It is true that such a situation may be all too common in the case of theories in the course of development, but in such cases one expects to find efforts being made to clear the situation up—not to find special principles being invoked to make a virtue of what seems to be necessity.

take place. The desired new atmosphere certainly requires that as a first stage, theories significantly different in their basic structure should be seriously discussed. At this first stage they need not be so successful as to be evidently correct. The conceptual muddle in the foundations of quantum theory is serious enough to make this first stage by itself well worth achieving.

The original shock to physicists produced by the quantum theory was that by the introduction of discrete structures into essentially continuous dynamics, advances which could not be ignored were made possible. The step was felt to be unscientific because scientific explanation demanded the possibility of indefinite refinement in the measurement of physical quantities, and this possibility, postulates of discreteness must necessarily restrict at some point. In other words a limit of any sort to measurement was felt itself to require explanation in a way that the absence of such limits does not: there is an asymmetry as regards the possibility of explanatory achievement between discrete and continuum theory.

Newer forms of quantum theory were looked upon to some extent as efforts to explain the novel quantum limitation on indefinite refinement of measurement (the Heisenberg principle has often been thought of in this way). However as the corpus of quantum theory grew it came to be thought that the problem of reconciling the quantum jumps with dynamics had disappeared from sight in its original form.

Now, however, it seems that we may be due for a return to a form of the old problem. In the papers and discussions that are printed in this book there appears a new preoccupation with finitism and discreteness and with systems of mathematics that take seriously discrete values only. In such systems it is likely that the only infinity that will be permitted will be a potential infinity—a finite set of points with a rule for adding more when they are needed, and the rule being defined by a physical operation that can actually be carried out.

This preoccupation with finitism can be traced back to a general agreement that there is something seriously wrong with the combination of quantum theory and continuity. This agreement in its turn depended on two or three major positions which emerged from papers and discussions, rather as a 'sense of the meeting'. First: the Bohr, or Copenhagen, interpretation, whether it is correct or not, is certainly insufficient, There has got to be considerable development beyond Bohr, as well as some reconsideration of assumptions.

Secondly: it is unlikely that there can be a unitary theory of physics which lays down the whole detail of nature, and there is no clearly marked line for future development that would lead to one complete theory. This conclusion may be held to follow from Bohr's views—and certainly for Bohr completeness would be a false ideal—but it seems in any case that quantum theory has roots in quite different soil from those of say, relativity. Thirdly: the massive coherence and invulnerability to change of ways of thinking and speaking which characterizes classical dynamics, was subjected to critical examination and reconsideration, but the importance which Bohr attached to it was not disputed.

It seems likely that the coherence of the classical picture has been responsible more than any other one thing for dogmatic assertions of completeness in physics and in science in general. This association of coherence with completeness was held to be misplaced, and the proper deduction to have been missed—namely that until one can achieve freedom from intuitive dependence on the classical picture one cannot handle discreteness.

Suppose for a minute that the complete picture of Bohr—that the classical language is inviolate and that discreteness must enter physics through the complementarity principle—has gone. Does this mean that this, the most sophisticated current theory which includes an 'epistemological' element in physics itself, was a mistake, and that the great physicists of the 'twenties were wrong in their intuition that quantum physics inescapably gave the observer a place within the theory? We shall not be able to answer this question until we know where the new theories that are to have a real place for the discrete locate that place.

REFERENCES

(1) Jauch, J. M., Wigner, E. P. and Yanase, M. M. Some comments concerning measurements in quantum mechanics, *Nuovo Cimento* (1967) **10**, 48.
(2) Born, M. *Atomic Physics* (Blackie, 1944).
(3) Blokhintsev, D. I. *Vvedenie v kvantovuiu mekhaniku* (Moscow and Leningrad, 1944). Quoted by L. R. Graham, *Slavic Review* xxv (1966) **3**, 395.
(4) Brillouin, L. *Science and Information Theory* (Academic Press, 1962).
(5) Temple, G. *Quantum Theory* (Methuen, 1931).
(6) Fock, V. *Dialectica* (1965), **19** 3/4, 237.

(7) Heisenberg, W. *Physics and Philosophy* (Allen and Unwin, 1958), 56.

(8) Birkhoff, G. and von Neumann, J. *Ann. Math.* (1936) **37**, 823.

(9) Dirac, P. A. M. *Quantum Mechanics* (Oxford), any edition.

(10) Dirac, P. A. M. *Quantum Mechanics* (Oxford), third edition, 8.

THE CONCEPTUAL PROBLEM
OF QUANTUM THEORY FROM THE
EXPERIMENTALIST'S POINT OF VIEW

O. R. FRISCH

I was given the pleasant job of saying welcome to everybody at the colloquium.

Papers had been circulated beforehand between the participants and in many cases these papers were modified as a result of our discussions to produce the material of this volume. The papers fell roughly into two categories. Some took quantum theory as being more or less correct but argued about how it should be interpreted; whether or not it is complete; how we should understand the process of measurement, and the famous reduction of the state vector and so forth; there were several papers concerned specifically with the question of probability which it was thought might serve for a subject for a special session.† Apart from that there are a number of papers which definitely assume that quantum mechanics has to be expanded or modified or even thrown away, and these are more or less covered by the word 'beyond' in the title of our colloquium.

I thought I would mention a few of the more specific problems in the first category. I know too little about the more imaginative attempts to extend quantum theory. It is a fact that classical physics on which many of us were brought up is much more specific and tends to be more complete in its description. It takes the line that there is definitely an outside world consisting of particles which have location, size, hardness and so on. It is a little more doubtful whether they have colour and smell; still, they are *bona fide* particles which exist there whether or not we observe them; they have momentum and position at the same time and if they have got an angular momentum it has got three components and so forth.

One of the arguments that has been going on for a long time is whether quantum mechanics is right in ascribing less detail to what it deals with—I am trying to avoid the word 'reality' or 'the thing' or anything like that—and there again one can take somewhat

---

† Several of the key probability theorists happened not to be able to come however, and so this aspect of the subject is not covered specifically in this book.

different attitudes. One attitude is to say that electrons are *there* but that although, in a sense, they are particles you must be careful not to ascribe position and momentum at the same time. Similarly you must be careful not to speak of more than one of the spin components.

The other attitude is to say that these things are not there; that they are merely pictures which we create in order to interpret the measurements; that measurements are done with macroscopic instruments consisting of a large number of atoms, and that for the description of instruments classical physics is adequate.

In fact some people strongly stress that it is an essential feature of an instrument that it should be described completely in classical terms. This has raised a question to what extent quantum mechanics can give an account of the actual process of measurement. The difficulty there is usually formulated by pointing out that the quantum-mechanical formalism describes a system—any system—including, for instance, a system that contains measuring instruments; that any system can be described by the state-vector or, possibly, by a mixture of state-vectors; and that in particular a microscopic system—a photon or electron—can be prepared in such a state that a particular state-vector gives a description of it. Let us say the particle has been prepared at a given time and position to be focussed on a certain point. If we then ask, what is its momentum, we get a confused answer, and we are assured that if we were actually to measure it and do that repeatedly we would get different replies.

That means that after that measurement has been performed the system is now in the new state which is described by a new state-vector. On the other hand the wave-equation says that the state-vector once given for the isolated system develops in a perfectly determined way and that there's no random element there. So if you get various answers to a question, what can be responsible for that variety? I understand that at present there exists a controversy, roughly speaking between a group of people which includes Wigner as the best known person and another group centred on Milan in Italy, and that these two have different views on how this reduction happens.

I do not want to enter into this because for one thing I do not understand it at all well. But perhaps one can make the difficulty clearer by speaking about an example such as Schrödinger's cat which I am sure you have all heard about. Schrödinger visualises a box containing an atom of radium, a counter, an explosive charge and a cat. When the radium atom decays it triggers the counter, the counter triggers the explosive, the explosive kills the cat. Then, according to some people—and I think Schrödinger himself said you must take this line if you take wave-mechanics seriously—after one half life this box contains neither a live cat nor a dead cat but a superposition of the wave-functions of both. To me this seems like a misuse of quantum theory.

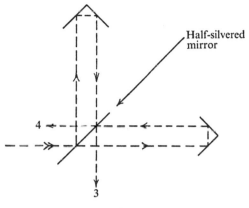

Fig. 1

I cannot quite put my finger on it but my feeling is that a cat should not be described by a wave-function or even by a super-position of wave-functions. But then the question is, where do you draw the line? What system do you allow to be described by state-vectors and when and where do you cease to do so? This, of course, can also be formulated by asking when an observation or measure-ment—I am using these terms interchangeably—has been made? In some of the earlier Copenhagen discussions one gained the impression that if for instance you measure the position of a particle by scattering a photon from it then the measurement has been made.

But if you reflect the photon afterwards by a suitably prepared set of mirrors and bring it back you might reverse what you have just done and thus undo the measurement.

One can think of a number of examples where a measurement appears to have been done but can, in fact, still be undone. I dis-cussed an example of one such thing in a paper some years ago, and it will help to illustrate some other points as well. Let us assume that I have a semi-reflecting mirror (fig. 1) and I let a light beam fall on it from the left. The light beam will then be split into two—one reflected and one transmitted. One can consider this as a wave phenomenon: that as the result of an incoming wave two outgoing waves are formed, each of an amplitude of $1/\sqrt{2}$ so that each carries half the intensity.

But can you also ask what happens to a photon in that case? In order to defend my view that you can ask that question legitimately, I have to show that we can make sure that one photon has hit the

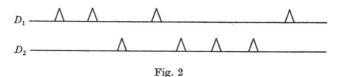

Fig. 2

mirror. There are various possible ways of doing it. For example if you want a beam of sodium light you can take a beam of sodium atoms and place next to it a suitable detector for sodium atoms, such as a hot tungsten wire. You illuminate the sodium beam with sodium resonance light. Every time an atom becomes excited and emits a photon to the right the atom itself jumps to the left. It will strike the wire which has been placed there and thereby trigger some electronic circuit to ring a bell to say the photon is on the way. Imagine that the photon is taken on a detour of a few hundred kilometres so that I have time to get ready for the photon to arrive. So I think there is nothing wrong with assuming that you can take a photon and know when it will fall on the mirror. What happens to that photon? Is it going to be reflected or transmitted or is there anything else? Of course the naïve view is that it gets either reflected or transmitted, and in support of that view you may say that you can place a detector D_1 in one beam and detector D_2 in the other beam and then you can arrange for both detectors to have their responses recorded as pulses on a moving strip of paper. Each time the bell rings you move your recording paper half an inch. Then if you get a recording like the one shown in fig. 2, you could verify that the photon always goes either the one way or the other way. I do not think that anyone has ever done that experiment or anything closely approaching it, but in principle it is possible to make detectors which are almost 100 per cent sensitive; so it could be done.

So far it all looks as if the photon went either one way or the other, but, of course, we all know that we could change our set-up a little bit by adding a few more mirrors and turning the whole thing into an interferometer (fig. 1). I use V-shaped mirrors to separate the outgoing beams and the returning beams; the two returning beams combine where they meet the mirror and at that point either we get a beam that goes down or one which goes back to where the light came from. Calling these beams 3 and 4, then, provided the length of the two paths are suitably chosen, we can arrange, for instance, that all the intensity should go into beam 3. That is so because we get interference, and that would seem to mean that the photon has become split and has travelled both paths.

I think that Bohr's attitude to this would have been to say that these are two different experiments. One must consider the experiment as a whole. If I place detectors into the two beams then I am looking for photons and I find photons. If I put no detectors into the beams but arrange the whole thing as an interferometer then I confirm the wave phenomenon as expected. But with the interferometer you cannot determine which way the photon is going. If you do that then you are no longer capable of observing interference, in other words then you get statistically the same number of photons recorded in beams 3 and 4.

One intriguing way that occurred to me (and that time I really thought that I had found something new) was to suspend the mirror from some strings in such a way that it can swing perpendicularly to its own plane. The nice point is that if you adjust the oscillation time of the mirror in such a way that it takes one half oscillation for the light to go out to the reflector and come back again, then you do not ruin the interference. Of course, the first thing that any visitor from Copenhagen will say is, 'Oh, yes. But if you suspend the mirror you ruin the interference because the position of the mirror is no longer well defined, not within one wave-length if you want to be able to measure the momentum of the photon. And in that case you can observe the momentum given to the mirror but you will get the same amount of light in the outward beams 3 and 4.'

However if you hang up that mirror in the way I described then you will still get interference because if the mirror happens to be by a certain amount out of its proper mid-plane at the moment that the photon strikes it then by the time the photon comes back it will be out by the same amount in the opposite direction. This just cancels the uncertainty in the light path; so the photon paths have the same length, and you do get interference between the two beams.

The trouble is that if you do that you still do not know which way the photon went. Should the photon be reflected on first meeting the mirror it pushes the mirror down, and coming back it goes through to 3 without giving momentum to the mirror. If one the other hand it goes through the mirror at first it will push it up later, on being reflected into beam 3. There seems to be a physical difference which would allow us to tell which way the photon went.

But that is not so. The point is that whether you push a pendulum to one side at one time or to the other side half a period later, you get the same result in the end! So, by observing the way the mirror swings after the photon has struck it, you do not learn anything.

On the other hand what I can do is that, after the photon has struck the mirror and is travelling out either along path 1 or path 2, I can decide to measure the momentum of the mirror. If I do that I have learned which way the photon went, but then, of course, I have ruined the interference.

The question to ask here is when is this measurement done? I am speaking now about the measurement which decides whether the photon has followed one or the other path, and my point is that it is not done as the photon becomes reflected, because if I do nothing else then this measurement will be undone again when the photon comes back. If it comes back, after it has been through, it will have given the same ultimate motion to the mirror irrespective of whether it followed path 1 or path 2.

It is not the reflection itself, therefore, which establishes the measurement. During the time the photon is out I can reflect an infra-red photon on the mirror and from the doppler shift determine the speed of the mirror; but there again is it the reflection of the photon which represents the measurement? I say it is not. Because once again, if I were fast enough, I could change my mind, send the infra-red photon back, return the momentum, whatever it was, to the mirror, and not have lost anything. So the outcome is that the measurement is not completed till something irreversible has happened, and that means that *irreversibility is an essential part of a measurement.*

I know this point is made in several of the papers and perhaps I should not labour it now. But what I want to stress is that a measurement is necessarily an interaction with an instrument, the irreversible behaviour of which is an essential part of its construction. One can argue for this in a very elementary way and say that measurement establishes information—knowledge—whatever you like to call it—and information is a commodity which exists after the observation has been made and then continues in the future unless it gets lost or destroyed. It defines by its very nature the direction of time. Information is something which is obtained at some time and continues into the future. Therefore, it is quite impossible to think in any way about measurement without including the existence of irreversible processes. This is what makes me sceptical about the possibility of describing the state of a measuring instrument, either before or afterwards, in terms of state-vectors.

I think I have laboured this point enough. Perhaps I should point out another way in which I think irreversibility comes in quite essentially, and that is something I have not seen mentioned much. In preparing an experiment you always make some fairly obvious assumptions. For instance, if you speak about a sodium beam you assume that the sodium atoms are in the ground state, as a rule. But how do you know they are in a ground state? You only know thanks to thermodynamics, thanks to the fact that they have been made in an evaporating device at a moderate temperature so that they are not excited. Similarly, when you use containers you assume the existence of solid bodies which are in thermal equilibrium at a fairly low temperature. I do not know what one can get out of that but I have a feeling that this is something which any theory of observation should take into account.

I think this is really all I wanted to say about the problem of measurement and the importance of the irreversibility. The example I have chosen here, a splitting mirror, is of course only one of a number of possible examples to be used. The problem is, in fact, more usually discussed in terms of two slits; a good many other people have chosen the Stern–Gerlach experiment as an example of an apparatus which makes a choice. There is, by the way, a difference here because in a Stern–Gerlach experiment, whether a silver atom goes into one beam or the other can be influenced by its previous preparation whereas with the splitting mirror we assume that it cannot. Otherwise much of the argument is similar.

The other thing I wanted to mention here is the fact that we tend not to take very seriously the feature of state-vectors or wave-functions that they are not necessarily three-dimensional. We usually think in terms of electrons or photons being described as point particles in three-dimensional space. Of course we are all quite willing to give them an extra degree of freedom for a spin, but in ordinary practical physics, at any rate, we are quite liable to forget that a system of two particles has got six dimensions of space (or more).

To dramatise this we have the Einstein–Podolsky–Rosen paradox. These three considered a system which breaks up into two particles. In classical parlance you can say that the two particles go off in opposite directions and with opposite speed. Therefore, if we measure either the speed or the position of one we know the speed or position of the other, and by deciding what to do with one particle we can decide what to learn about the other one, which may be far away.

This paradox can be sharpened by considering other degrees of freedom, and my own favourite example is the decay of positronium into two photons. There it was shown that the two photons must be polarised at right angles to each other. At first, when people tried to test this it was thought this meant that if one photon happens to have, shall we say, vertical polarisation, the other one must be horizontally polarised; if I pick one of them up through a polaroid in the vertical direction the other one will with certainty go through a polaroid with horizontal polarisation. But what happens if you turn both polaroids by 45°? You would say that's bad luck for both photons: one of them has a 50 per cent chance to go through; and the other one knows nothing about it and thus again has a 50 per cent chance of going through.

In fact a more careful study shows that this is not so. Quantum theory predicts that whenever you put the two polaroids in any arbitrary direction, at right angles to one another, if one photon was to go through one of them, then the other photon would be sure to go through the other one! This cannot be described in any conceivable way by giving each photon a certain polarisation; we must consider the two photons as being one system. Or if you prefer it, you may say that one photon arriving at its polaroid, if it goes through, telegraphs back—with speed exceeding that of light—to tell the other photon 'go through your polaroid'; or 'please adopt the polarisation corresponding to the one which has just been measured for me'.

How should we think about this? Should we simply accept that quantum theory gives a separate space of three dimensions for each particle, and additional dimensions for each degree of freedom of polarisation? Or should we say that space is three-dimensional, but that particles are in collusion, that they, as it were, communicate in some way or another over great distances? (Incidentally if you try to work out a signalling system with more than the speed of light on this basis it does not work. I have tried it.) This does not look like a very good description. At the same time I am aware that Feynman and Wheeler in their famous paper make use of a kind of signalling with the reverse speed of light ('into the past'). Whether that can be used in the present context I do not know. I am not here to give answers, I am here to throw out questions.

If you accept that each particle has its own space then why do we labour under the delusion that we are living in a three-dimensional space? Somehow or other this very complex multidimensional system must fall into the classical semblance of 3-space when you consider sufficiently many particles and are in a position to ignore their correlations, as you usually can.

These are, I think, the only questions which I wanted to bring up. Some of the papers, as I have said, bring up questions which are much further-going; they query the structure of space and time: they suggest using a discretum instead of a continuum, and so forth. It would only be wasting time to discuss all this now. I hope I have set the ball rolling and not merely kicked it off in a random direction.

II

NIELS BOHR AND COMPLEMENTARITY: THE PLACE OF THE CLASSICAL LANGUAGE

No one came to the colloquium, even from Copenhagen, prepared without reservation to defend the position of Niels Bohr on the foundations of quantum theory. Almost all participants, on the other hand, were concerned to establish ways in which an advance on Bohr's position might be secured, and it seemed appropriate that a photograph of Bohr should watch us with knitted brows from the mantelpiece of the King's College Audit Room where we met.

We print short papers by von Weizsäcker and by Bohm who explicitly discuss the background of Bohr's thinking. Later in this volume both these authors are to propose views which go beyond what would have been acceptable to Bohr, though both would claim that an understanding of Bohr's view is a prerequisite for subsequent modification of it. In this estimate of the Bohr legacy, von Weizsäcker and Bohm are really typical of the majority of participants in the colloquium who had each come to terms in some way with the massive coherence and pervasiveness of the language and way of thinking of classical physics, as well as with the apparently irreducible epistemological element inside quantum theory. The complementarity doctrine of Bohr comprehended both of these important aspects of the contemporary physics picture, and it seems that any doctrine which is to have a chance of superseding that of Bohr will have to do at least as well in achieving this synthesis, whatever else it achieves. In this way we appear to be assured that the profundity of vision of Bohr in physics cannot indefinitely be lost to sight or trivialized by merely technical development.

THE COPENHAGEN INTERPRETATION

C. F. VON WEIZSÄCKER

I come to praise the quantum, not to bury it.

F. BOPP

One can think of two reasons for wishing to go beyond quantum theory: dissatisfaction with the present status of the theory, and a general belief in further progress. I shall treat the first reason in this section, the second in my paper in Part VI of this book.

There is no known reason to be dissatisfied with the success of quantum theory, with the possible exception of the difficulty in reconciling it with relativity in a quantum field theory of interacting particles. This difficulty, however, belongs to an unaccomplished theory, and physicists may have good reason to believe that it will disappear with the completion of elementary particle physics; I shall briefly return to it in section 4 of my paper in Part VI. Thus the mathematical formalism of quantum theory and its usual mode of application to practical problems do not seem to need improvement.

On the other hand, there has been considerable discontent during the last decades with what is generally called the Copenhagen interpretation of quantum theory. I would prefer to call it the Copenhagen interpretation of the formalism, an interpretation by which the formalism is given a sufficiently clear meaning to become part of a physical theory. In so saying I express the opinion that the Copenhagen interpretation is correct and indispensable. But I have to add that the interpretation, in my view, has never been fully clarified. It needs an interpretation itself, and only that will be its defence. This situation, however, should not be surprising once we realize that we are here concerned with the basic problems of philosophy. To use a simplifying example: one should not expect it to be easier to understand the atom than to understand number; and the battle between logicism, formalism, and intuitionism about the meaning of number is so far by no means decided.

In the present paper I cannot treat this question within its full philosophical horizon. I shall only point out what I consider to be the crucial point in the Copenhagen interpretation. The point is envisaged, but not very luckily expressed, in Bohr's famous statement that all experiments are to be described in classical terms. This statement, if correctly interpreted, is, I maintain, true and essential; its main weakness is the insufficient elucidation of the meaning of the word 'classical'. Its clarification is difficult yet necessary because Bohr's statement implies an apparent paradox: classical physics has been superseded by quantum theory; quantum theory is verified by experiments; experiments must be described in terms of classical physics. The paradox will only be resolved by stating it as poignantly as possible.

It will serve the clarification if we first say that Bohr was not a positivist, and why. A few years ago E. Teller reminded me of two remarks Bohr made when we were staying together at his institute. One of them was made after Bohr had addressed a congress of positivist philosophers. Bohr was deeply disappointed by their friendly acceptance of all he said on quantum theory, and he told us: 'If a man does not feel dizzy when he first learns about the quantum of action, he has not understood a word.' They accepted quantum theory as an expression of experience, and it was their *Weltanschauung* to accept experience; his problem, however, was precisely how such a thing can possibly be an experience.

The Copenhagen interpretation is often misrepresented both by some of its adherents and some of its adversaries as stating that what cannot be observed does not exist. This is logically inaccurate. What the Copenhagen interpretation needs is only the weaker statement: 'what is observed, certainly exists; about what is not observed we are still free to make suitable assumptions'. This freedom then is used to avoid paradoxes. Thus Heisenberg's discussion of thought experiments with respect to the uncertainty principle is no more than a refutation of an accusation of inconsistency. If we accept the formalism of quantum theory in its usual interpretation we have to admit that it does not contain states in which a particle both has a well defined position and a well defined momentum; the exclusion of these states seems† essential in order to avoid contradictions with the probability predictions of the theory. The objection is raised that both position and momentum can be measured, hence they exist. This objection is countered by the statement that, *if* quantum theory is correct, position and momentum cannot be measured at the same time, and hence the defender of quantum theory cannot be forced to admit that they exist at the same time, that is in the same state of the

---

† I here leave aside the question whether the contradictions might be avoided by theories like Bohm's. We are at this point concerned with the consistency and not with the uniqueness of the Copenhagen interpretation.

particle. Thus in Heisenberg's view it is only the positive knowledge of the state of particles, in quantum theory, that excludes the rejected states; states which then, quite deservedly, also turn out to be un-observable. Heisenberg made the simple, but relevant comparison that we did not reject the theory that the earth's surface is an infinite plane just because we are not able to go beyond certain bounds, but because we are fully able to go around the earth and thereby to prove that it is a sphere. The fact not expressed in this comparison is, how-ever, that quantum theory does not reject particular models but the very concept of 'models'.

These last remarks on the difference between the Copenhagen interpretation and positivism were logical and negative. In an epi-stemological and positive manner the difference is formulated in Bohr's statement on classical concepts. At least the more naïve posi-tivist schools have held that there are such things as sense data, and that science consists in connecting them. Bohr's point is simply that sense data are not elementary entities; that what he calls phenomena is given, rather, only in the full context of what we usually call reality— a context which can be described by concepts; and further, that these concepts fulfil certain conditions which Bohr took to be characteris-tic of classical physics. Here the second Teller anecdote may be in place. Once at afternoon tea in the Institute Teller tried to explain to Bohr why he thought Bohr was wrong in thinking that the historical set-up of classical concepts would forever dominate our way of expressing our sense experience. Bohr listened with closed eyes and finally only said: 'Oh, I understand. You might as well say that we are not sitting here, drinking tea, but that we are just dreaming all that.'

Evidently Bohr expressed his point here without proving it. What are the conditions to be fulfilled by sense experience, and why should they be so unalterable?

Bohr used to speak of two conditions: space–time description, and causality. They go together in the objectivated model of events offered by classical physics, while they are broken asunder by the discovery of the quantum of action, being reduced to complementary means of description. If a physical system is to be used as an instru-ment of measurement, it must however both be describable in the space and time of our intuition and be describable as something functioning according to the principle of causality. The first condi-tion ensures that we are able to observe it at all, and the second that we can draw reliable conclusions from its visible properties (like the position of a pointer on a scale) to the invisible or dimly visible properties of the object which we observe by it. If Bohr is right in saying that space–time description and causality go together only in classical physics, his view that a measuring instrument must admit of a classical description seems to be inevitable.

Before asking the next question I should like to point out how this analysis distinguishes Bohr's view from positivism. To him what can be described classically is a 'thing' in the common-sense meaning of the word. The fact that classical physics breaks down on the quantum level means that we cannot describe atoms as 'little things'. That does not seem to be very far from Mach's view that we should not invent 'things' behind the phenomena. But Bohr differs from Mach in maintaining that 'phenomena' are always 'phenomena about things', because otherwise they would not admit of the objectivation without which there would be no science about them. The true rôle of things for Bohr is precisely that they are not 'behind' but 'in' the phenomena. This is very close to Kant's view that the concept of an object is a condition of the possibility of experience; Bohr's dichotomy of space–time description and causality corresponds to Kant's dichotomy of the forms of intuition and the categories (and laws) of thought which make experience possible only in conjunction. The parallelism of the two views is the more remarkable since Bohr seems never to have read much of Kant. Bohr differs from Kant in having learnt the lesson of modern atomic physics, which taught him that there can be science even beyond the realm in which we can meaningfully describe events by properties of objects considered independent of the situation of the observer; this is expressed in his idea of complementarity. Bohr differs far more radically from Mach in denying that there are sense data as distinct from what can be observed in things. Since his positivist audience did not understand this, they disappointed Bohr by easygoing acceptance of the quantum of action. It is typical of this attitude that it has no real use for the idea of complementarity.

The next question is how Bohr knew that our space–time intuition was unalterable and that space–time description and causality go together only in classical physics. My proposed answer is that Bohr was essentially right but that he did not know why. This is, I think, indicated in his ways of expression. He never stated in so many words that space–time intuition cannot change and that the set of classical laws (say Newton's mechanics and Maxwell's electrodynamics) comprised what was necessary and sufficient for objectivation;

yet he expressed himself in a manner that would seem indefensible if these two statements were wholly wrong. I should like to offer an hypothesis of my own on the question.

Classical physics is a very good approximation but not an exact description of phenomena. Perhaps the statement of my hypothesis should be prepared by stressing that not only is classical physics refuted empirically but that it has or had a very scanty *a priori* chance of being an exact theory of phenomena, at least if it is to embrace both thermodynamics and an account of the continuous motion of continuous bodies. Planck's ultraviolet catastrophe would probably have presented itself in any sufficiently elaborate classical thermodynamics of continuous motion. (I think, by the way, that this difficulty would turn up in any attempt to undermine quantum theory by a classical theory of hidden parameters.) Having thus to accept the essential falsity of classical physics, taken literally, we must ask how it can be explained as an essentially good approximation. This amounts to asking what physical condition must be imposed on a quantum-theoretical system in order that it should show the features which we describe as 'classical'. My hypothesis is that this is precisely the condition that it should be suitable as a measuring instrument. If we ask what that presupposes, a minimum condition seems to be that irreversible processes should take place in the system. For every measurement must produce a trace of what has happened; an event that goes completely unregistered is not a measurement. Irreversibility implies a description of the system in which some of the information that we may think of as being present in the system is not actually used. Hence the system is certainly not in a 'pure state'; we will describe it as a 'mixture'.† I am unable to prove mathematically that the condition for irreversibility would suffice to define a classical approximation, but I feel confident that it is a necessary condition.

If the hypothesis is correct it might show that Bohr, far from

† Wigner has objected against the description of a measurement by a mixture that a unitary transformation of a mixture into another mixture (which would be quantum-theoretically admissible) cannot increase the entropy and hence cannot describe the irreversible traits of the measuring process. That is correct, but it is no objection. Even in classical physics an increase of entropy cannot be described as an 'objective' event. If we stir up an incompressible fluid which is coloured half white and half red (Gibbs) we do not produce pink areas but only complicated boundaries of white and red areas. But when the boundaries have become sufficiently complicated, then *we* can no longer distinguish them and hence we see pink. Correspondingly an irreversible process cannot be described in quantum theory by a unitary transformation of mixtures; our description must rather 'jump' into a mixture with a higher entropy.

stating a paradox, rather stated a truism, though a philosophically important one: a measuring instrument must be described by concepts appropriate to measuring instruments. It is then not unnatural further to assume that classical physics, in the form in which it developed historically, simply describes the approximation to quantum theory appropriate to objects as far as they really can be fully observed. If we accept this we have to conclude that, at least to the extent that quantum theory itself is correct, no further adaptation of our intuitive faculty to quantum theory is needed or possible. A mind that observes nature by means of instruments themselves described classically cannot possibly adapt to the actual laws of physics (i.e., the quantum laws) other than by describing nature classically. To ask for another description of phenomena would be to ask for an intrinsic impossibility—unless quantum theory is wrong after all.

This goes as far as I feel able in interpreting the Copenhagen interpretation. One would certainly wish to understand why nature should, of all things, obey the laws of quantum theory which were accepted in this section as given. This will be one of the guiding questions of my later contribution. To end the present discussion I explicitly formulate a methodological principle that I have used in it implicitly. I propose to call it the principle of semantic consistency.

A mathematical formalism like, e.g. Hamilton's principle with its mathematical consequences, Maxwell's equations with their solutions, or Hilbert space and the Schrödinger equation, is not *eo ipso* a part of physics. It becomes physics by an interpretation of the mathematical quantities used in it, an interpretation which one may call physical semantics. This semantics draws on a previous, though incomplete knowledge of the phenomena which we hope to describe more accurately by the formalized theory. Thus we previously knew what we mean by a body, a length, or, for the later theories, by a field of force or an observable like energy, momentum, etc. Since the theory ascribes mathematical values to quantities like length, force, momentum, it is indeed necessary that we possess previous knowledge about the practical methods of measuring those quantities. But then in many cases the measuring instruments will themselves virtually be objects of a description by the theory. This possibility submits the whole theory (that means the formalism plus its physical semantics) to an additional condition, precisely the condition of semantic consistency: The rules by which we describe and guide our measurements, defining the semantics of the formalism, must be in accordance with the laws of the theory, that is with the mathematical statements of the formalism as interpreted by its physical semantics.

It is by no means trivial that a new theory should automatically fulfil this condition. Thus Einstein's famous analysis of simultaneity was an analysis of the semantic consistency of a theory that obeyed the mathematical condition of Lorentz invariance; it showed that space and time had to be reinterpreted in order to reconcile everyday experience with the theory. The quantum theory of measurement aims at a proof of the semantic consistency of general quantum theory. So far no theory in physics has ever been fully submitted to the test of semantic consistency, for all theories accept certain phenomena as given which they do not explicitly describe. Thus quantum theory accepts that there are objects which have the nature of particles, a fact which perhaps in the end will be explained by elementary particle theory.

ON BOHR'S VIEWS CONCERNING THE QUANTUM THEORY

D. BOHM

The views that I express in my main contribution to this colloquium later on are in certain ways rather similar to those of Bohr, while in other ways, they are basically different. It will perhaps be useful here to go into Bohr's notions in some detail, in order to bring out what these similarities and differences are.

First, it should be said of Bohr that his writings show an unusually strong emphasis on *coherence and consistency of language*. In this regard, one of his most important contributions is that he saw, at least implicitly, that the *form* of a coherent communication has to be in harmony with its *content*. For example, if someone shouts 'Be quiet!' the noisy, angry and turbulent form of such speech excites and stirs up the hearer; and this evidently interferes with the communication of the intended content (which is the need for a situation of peace and calmness). What has generally not been noticed is that the way in which physicists have usually discussed the quantum theory has a rather similar disharmony of form and content, which tends to lead also to confusion. Bohr's writings are characterized by a highly implicit and carefully balanced mode of saying things, which makes reading his work rather arduous, but which is in harmony with the very subtle content of the quantum theory.†

One of the major sources of disharmony, and therefore confusion, in most discussions of the content treated in the quantum theory is that there is a strong tendency to continue to use a basic *form* of discourse that was appropriate only in classical physics. This form is that of describing the world as a union of disjoint elements. Thus, most physicists (especially those who follow along lines initiated by von Neumann) continue to talk about a 'quantum system' as if it were constituted of interacting components (e.g. particles) which exist separately from each other, and from the instrument that is used in 'observing the quantum state of the system'. What is most characteristic of Bohr is that he does not use such a language form at all. Indeed, his discussion implies that this mode of description is irrelevant, in the quantum context. What is relevant instead is the

† For a more detailed discussion of this question, see D. Bohm and D. Schumacher.[1]

wholeness of the form of the experimental conditions and the content of the experimental results† (from which it follows that it will not be consistent in this context to talk in terms of the classical notion of a union of disjoint elements).

In order to bring out Bohr's point of view in more detail it will be useful to discuss Heisenberg's well known microscope experiment, but in a rather different way from that which has generally been adopted. To do this, one can begin by asking in terms of *classical physics*, what it means to make measurements of position and momentum. We start, however, not as is usually done, with a *light* microscope but, rather, with an *electron microscope*.

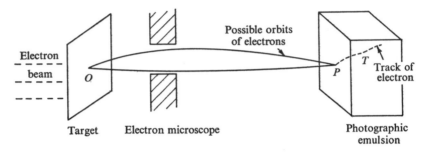

Fig. 1

In fig. 1 there is in the target an 'observed particle' at O, assumed to have initially a known momentum (e.g. it may be at rest, with zero momentum). Electrons of known energy are incident on the target and one of these is deflected from the particle at O. It goes into the electron microscope, following an orbit that leads it to the focus at P. From here, the particle leaves a track, T, in a certain direction, as it penetrates the photographic emulsion.

Now, the *directly observable results* of this experiment are the position P, and the direction of the track, T. But, of course, these in *themselves* are of no interest. It is only by knowing the *experimental conditions* (i.e. the structure of the microscope, the target, the energy of the incident beam of electrons), that the experimental results can become relevant in the context of a physical inquiry. With the aid of an adequate description of these conditions, one can use the experimental results described above to make *inferences* about the position of the 'observed particle' at O, and about the momentum transferred

† This question is discussed extensively in reference (1), as well as in a talk for the Illinois Symposium on the Philosophy of Science.[2]

to it in the process of deflecting the incident electron. As a result, one 'knows' both the position and the momentum of this particle at the time of deflection of the incident electron.

All this is quite straightforward in *classical* physics. Heisenberg's novel step was to consider the implications of the 'quantum' character of the electron that provides the 'link' between the *experimental results* and *what is to be inferred from these results*. The electron can no longer be described as being just a classical particle. Rather, it also has to be described in terms of a 'wave' as shown in fig. 2. Electron

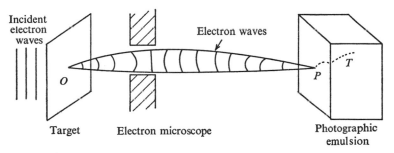

Incident electron waves

Electron waves

O

T

P

Target   Electron microscope   Photographic emulsion

Fig. 2

waves incident on the target are said to be diffracted by the atom at *O*, then pass through the microscope, where they are also diffracted, and are focussed at the photographic emulsion. Here, they are said to determine only the 'probability' that a track *T*, begins at *P*, and goes in a certain direction.

However, as was implicit in Heisenberg's discussion (and as was later brought out more explicitly by Bohr), this whole situation involves something radically new and not 'coherent' with classical notions. Heisenberg tried to express this novelty by saying that both the position *x* and momentum *p* of the 'electron link' between *O* and *PT* are 'uncertain', the extent of this uncertainty being measured by the 'uncertainty relationship' $\Delta x \Delta p \geqslant h$. But this involves a very significant kind of disharmony between the form of the language and the content to which Heisenberg implicitly intended to call attention. [1], [2] The form of the language implies that the 'link' electron actually has a definite orbit that is, however, not precisely known to us. Bohr [3] gave a thorough and consistent discussion of this whole situation, which made it clear that the orbit of the electron is not 'uncertain' but, rather, that it is what he called *ambiguous*. Unfortunately, even this word does not give a very clear notion of what is meant here. Perhaps one could say instead that both the notion of a particle following a well defined orbit (whether known or unknown) and of a wave following a similarly well defined wave equation are not *relevant* to the 'quantum' situation. What we have to deal with here is a radically new form of description, that is incompatible with either of the old forms. [1], [2]

Now, because of the irrelevance in the 'quantum' context of the description of the 'electron link' in terms of well defined particle orbits or in terms of well defined wave motions, it followed that from

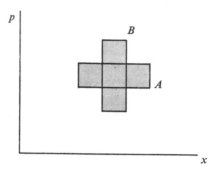

Fig. 3

the observed results of an experiment, one could no longer make inferences of unlimited precision about the observed object. But something more also followed, the very deep and far reaching significance of which most physicists tended to overlook. To see what this is, we note that from a particular set of experimental conditions, as determined by the structure of the microscope, etc., one could in some rough sense say that the limits of relevance of the classical description of the 'observed object' are indicated by a certain cell in the phase space of this object, which we denote by A in fig. 3. If, however, we had different experimental conditions (e.g. a microscope of another aperture, electrons of different energy, etc.), then these limits would be indicated by *another* cell in this phase space, indicated by B in the diagram. Both cells would have the same area, h, but their 'shapes' would be different.

Now, in the corresponding discussion of the classical situation, it is possible to say that the experimental results do nothing more than permit inferences about an observed object, which exists separately and independently, in the sense that it can consistently be said to 'have' these properties whether it interacts with anything else (such as an observing apparatus) or not. A description of the experimental conditions is needed to permit these inferences to be carried out properly, but this description is in no way needed for saying what is meant by the properties of the observed object, i.e. position and momentum.

However, in the 'quantum' context the situation is very different. Here, certain relevant features of what is called the observed particle, i.e. the 'shapes' of the cells in phase space, cannot properly be described except in conjunction with a description of the experimental conditions. Nor can one say the 'shapes' correspond only to our lack of knowledge about the precise position and momentum of the observed object, considered as separate and disjoint from the overall experimental arrangement. Indeed, as a more extensive discussion of the mathematical formalism shows, the region in which 'the wave function of the observed object' (and its Fourier coefficients) are appreciable corresponds to the 'shape' of the cell, as discussed above. But this wave function is co-ordinated, or 'correlated', to that of the 'link electron' in such a way that one has no meaning without the other. As a result, 'the wave function of the observed object', which gives the fullest possible formal means of determining averages of physical properties, cannot be discussed relevantly, apart from the experimental conditions, which provide a necessary context for a treatment of the wave function of the 'link electron'.

Thus, the mathematical formalism contains a reflection of the general situation with regard to 'shapes' of cells in phase space that has been described here in informal terms. Therefore, the description of the experimental conditions does not drop out as a mere intermediary link of inference, but remains amalgamated with the description (both formal and informal) of what is called the observed object. This means that the 'quantum' context calls for a new kind of description, which does not make use of the potential or actual separability of 'observed object' and 'observing apparatus'. Instead, as has been indicated earlier, the form of the experimental conditions and the content of the experimental results have now to be one whole, in which analysis into disjoint elements is not relevant.

The irrelevance of such an analysis (e.g. in terms of the notion that the 'observed object is disturbed by the observing apparatus') is brought out in a very forceful way in the famous discussion between Einstein, Podolsky and Rosen [4] and Bohr. [5] This showed that such wholeness of description was needed even when observations are carried out very far from each other in space and under conditions in which one would have said in terms of classical physics that no mechanical or dynamical contact or interaction between these observations is possible. [1], [2]

In a context in which the detailed description of the 'observed object' is not relevant, so that the 'cells' in phase space could be replaced by points, however, the 'shapes' of the cells would not matter. Thus, in such a context, all significant aspects of the 'observed object' could meaningfully be described without bringing in a description of the experimental conditions. Thus, one could consistently use the traditional notion that the object can be discussed in terms of a 'state' or of properties, that need not refer in any essential way to anything else at all. Therefore, the description in terms of a potential or actual disjunction between observed object and observing apparatus is a relevant simplification, in a context in which the fine details of the 'quantum' description do not matter. But where the details are significant, this simplification cannot properly be carried out, and one has to return to a consideration of the wholeness of the total experimental situation.

What is meant here by wholeness could be indicated in a somewhat informal and metaphoric way by calling attention to a pattern (e.g. in a carpet). In so far as what is relevant *is* the pattern, it has no meaning to say that different parts of such a pattern (e.g. various flowers or trees that are to be seen in the carpet) are disjoint objects in interaction. Similarly, in the 'quantum' situation, terms like 'observed object', 'observing instrument', 'experimental conditions' and 'experimental results' are just aspects of a single overall 'pattern' that are, in effect, abstracted and 'pointed out' or 'made relevant' by our mode of discourse. Thus, it has no meaning to say, for example, that there is an 'observed object' that interacts with the 'observing instrument'.

Thus far, Bohr's views are in general harmony with those adopted in the discussion of 'hidden variables', as given in my article which appears later. But now we come to an important difference between Bohr's views and my own. For Bohr went on to say that the terms of discussion of the experimental conditions and of the experimental results were *necessarily* those of 'everyday language', suitably 'refined' where necessary, so as to take the form of classical dynamics. It was apparently Bohr's belief that this was the only possible language for the *unambiguous communication* of the results of an experiment. He saw moreover that this kind of communication could be consistent with all that is meant by the term 'quantum' only if it were further ruled that the experimental conditions needed for the precise definition of one of a pair of 'complementary' variables (e.g. position) are not compatible with those needed for the precise definition of the other member of the pair (e.g. momentum). Since in classical physics *both* are required for the prediction of the future behaviour of a 'system', it follows that classical theory is no longer valid as a means of making *inferences* of this kind. Rather, it is now only a source of terms of *description of the experimental phenomena* (e.g. 'position', 'momentum', etc.).

Both the experimental conditions and the experimental results are, as in classical physics, to be described in terms of position and momenta of various objects and parts of objects and parts of objects that make up the total experimental arrangement. But (as indicated in the discussion of the microscope experiment) the connections between these conditions and results and the inferences that are to be drawn from them must now be obtained from the quantum algorithm, with its purely statistical interpretation. So, in a certain sense, Bohr takes quantum theory to be a kind of 'generalization' of classical theory. What is to be observed is always described in classical language (regarded here as a refinement of ordinary 'every-day' language), but the generalization consists in replacing the classical algorithm (the differential equation applying to individual systems) by a quantum algorithm (matrix theories applying only to statistical ensembles).

Because the experimental facts actually did disclose a statistical fluctuation of results, in accordance with what was to be inferred from the quantum algorithm, Bohr concluded that his general mode of using language was indeed relevant to the facts that were available to him at the time. But he went further than this. Since the classical language was supposed to be the only possible means of unambiguous communication, and since the terms of this language could not consistently be defined together, if one used the quantum algorithm for making statistical inferences in the usual way, he also concluded that it is impossible to find *any unambiguous language at all* that could treat the order of occurrence of these statistical fluctuations as relevant. Therefore it would necessarily be a source of confusion merely to entertain the notion of 'hidden variables' in terms of which these contingent fluctuations would be revealed unambiguously in a new field of novel orders of necessity. In this way Bohr was led to the conclusion that the 'quantum' implies absolute contingency, i.e. the necessity for 'complete randomness' in the results of what are called individual experiments.

On the other hand, there is a consistent way of doing just what Bohr would have said to be meaningless and irrelevant, in a theoretical language form that has a certain similarity to that of Bohr, and I shall describe this in my later paper.

A key difference between these forms is, however, that in terms of the language of my paper, one can discuss a possible new kind of significance for the *order* of successive operations (i.e. 'measurements'). According to current quantum theory, this order has to be 'random'. Indeed, there is no room, either in the formalism, or in the informal language of the theory, even to talk about a 'non-random' order of this kind. Moreover, as one can see, the particular order determined by the contingent parameters (i.e. 'hidden variables') also could not be incorporated into any classical theory. Thus, if this order is significant, one will have to describe the *experimental results themselves* (and more generally, the experimental conditions, as well) in terms of a new language form that is neither 'classical' nor 'quantum'.

What is called for, in my view, is therefore a movement in which physicists freely explore novel forms of language, which take into account Bohr's very significant insights, but which do not remain fixed statically to Bohr's adherence to the need for classical language forms, limited by the quantum algorithm, in the description of the experimental conditions and the experimental results.

In my article on 'hidden variables' (as well as in references (1) and (2)) some preliminary steps of this kind have been suggested. It seems to me that the work done thus far indicates clearly that there is a considerable scope for such exploratory experimentation with new language forms, and that such experimentation shows genuine possibilities for being fruitful.

REFERENCES

(1) Bohm, D. and Schumacher, D. On the role of language forms in theoretical and experimental physics. Also: On the failure of communication between Bohr and Einstein. (Preprints.)

(2) Bohm, D. Science as perception communication (Talk to Illinois Symposium on the Philosophy of Science, 1969).

(3) Bohr, N. *Atomic Theory and the Description of Nature* (Cambridge University Press, 1934). Also *Atomic Physics and Human Knowledge* (New York: Wiley, 1958).

(4) Einstein, A., Podolsky, B. and Rosen, N. *Phys. Rev.* (1935) **47**, 777.

(5) Bohr, N. *Phys. Rev.* (1935) **48**, 696.

III

THE MEASUREMENT
PROBLEM

The measurement problem is the problem of incorporating the measurement—or observation—process within the deductive structure of quantum theory. To solve the measurement problem in this sense is seen by some physicists as all that is needed to conquer the basic conceptual difficulty of quantum theory and this section contains papers by writers who hold this view and by some of their critics.

The critics (who happened to be in the majority) split into very different groups, but all of them would argue that one cannot reasonably expect to obtain a detailed analysis of the measurement process using quantum theory itself, because it is a postulate of quantum theory—in the opinion of these critics—that a detailed analysis of the actual observation process is impossible. Of these critics, some are satisfied with quantum theory as it stands, and some are not. Those who are satisfied accept the quantum theoretical formalism including a way of specifying the observation process specified by postulates which are algorithmically complete but which necessarily defy comprehension in the classical manner. From this point of view, to turn the quantum-mechanical machinery onto the measurement process is to operate in a circle: one possesses the quantum-mechanical machinery only because one has abandoned the expectation of a detailed dynamics of the observation process.

Paradoxically, many of the critics of the programme of the measurement theorists who want to go beyond existing quantum theory hold that position because—in common with the measurement theorists— they are not satisfied with the quantum theoretical account of observation as it stands at present. In a way they too, therefore, believe that there is a measurement problem, but they think that its solution can only lie outside the quantum theory.

The measurement theorists themselves exhibit quite a range of positions. Prosperi and his colleagues expect quantum theory—like any physical theory—to apply universally and therefore suppose that its detailed application to the measurement process can only be a matter of technique. In the discussion there was some criticism of this position on the ground that there was in fact no need to demand this kind of universality of physical law. (The term 'completeness' was used in the colloquium to cover this classically familiar idea of physical law as applying universally.)

It was unfortunate that the elucidation of the foundational issues that have just been mentioned interfered with the discussion of technical details that confront the measurement theorist, though Bub's paper is specifically directed at the more general of Prosperi's arguments.

Groenewold's position was different. He was not satisfied to ignore all the problems surrounding the observation idea (including those of complementarity) and to work with quantum theory as a self-contained structure as did Prosperi. He looked for a development of the quantum theoretical formalism to systems of many degrees of freedom that would exhibit complementarity and otherwise provide a detailed picture of the deep quantum theoretical doctrines that surround 'observation'.

Garstens surveys the literature on the measurement problem, as well as the points of view represented in the colloquium on it, to see how far an adequate bridge between microscopic and macroscopic physics exists. With Prosperi he starts from statistical mechanics, but he thinks that the much-discussed 'collapse of the wave function' which has to be assumed by quantum theory poses difficulties which have to be discussed at a foundational level and cannot be resolved at the level of technical quantum theoretical formalism.

Whiteman's paper, which ends this section, is a discussion of the observation process and of the collapse of the wave function for which he imports new ideas into quantum theory based on a detailed phenomenological analysis. His treatment of detailed cases compels one to realize how stylized much discussion about observation in quantum theory has become, in spite of the claim to phenomenological orientation of quantum theory.

QUANTAL OBSERVATION IN STATISTICAL INTERPRETATION

H. J. GROENEWOLD

There are four points I want to mention in this paper. The first one is about the technical meaning and the practical use of the wave function Ψ or more generally the statistical operator $\hat{k}$. As a simple example let me consider a Young experiment in which an incoming beam of photons or other particles is passing through two slits in a screen S and gives an interference pattern in the x-direction on say a scintillation screen R (fig. 1a). First we make the intensity so small, that particles only now and then strike the screen. Then we replace

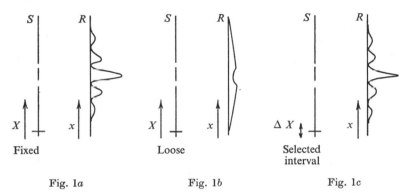

Fixed   Loose   Selected interval

Fig. 1a   Fig. 1b   Fig. 1c

the scintillation screen by a series of photographic plates, so that we catch each particle on a fresh plate. At the end we have a very large number of plates. We call it an ensemble. If we put all the plates together, we see just the original interference pattern.

Now we free the screen S with the two slits, so that it can move in the X-direction. Then instead of the interference pattern we get just the two diffraction patterns of the two slits as if the two outcoming partial beams were incoherent (fig. 1b). You might do this by suspending the screen on a string with a pulley and a counterweight. You do nothing to it, but there is an exchange of momentum in the vertical X-direction with the beam particles. You do not know either the momentum or the position of the screen, but you can easily work out that the interference pattern becomes smeared out. Even after the particle has gone through, I can decide to measure for instance

43

the position X of the screen S. If I do that after every passage of a particle, and then select all the plates for which X lies within a certain sufficiently small interval ΔX and put these plates together, I get back the interference pattern (fig. 1c). In this way the whole ensemble can be divided into subensembles, which each show an interference pattern, though each at a different height. This is only to show by an illustration that it has no meaning to ask in the case of a single plate whether we have coherent or incoherent beams coming through the two slits. The concepts of coherence and incoherence can only be applied to an ensemble or subensemble with a very large number of samples and not to a single individual sample. This is not only true of the concept of coherence, but also of the concept of the wave function or the statistical operator. What I should like to advocate is a statistical interpretation which says that if we are talking in terms of statistical operators or of wave functions, we can only do so correctly if we consider an ensemble and not if we consider a single sample. If you do insist on considering individual samples, you have to adopt some other interpretation, for instance in terms of hidden parameters. The statistical operator is not sufficient to meet such a detailed description.

So we have two interpretations: the statistical and the individual one. I shall consider the statistical interpretation as a legal and consistent procedure in the approximation of non-relativistic quantum mechanics. If we could keep our communication canals open here, we might perhaps come to an agreement, but it would involve us in some very exhausting discussions. About the individual interpretation the situation is much more difficult. Personally I do not believe that an individual interpretation is possible at all, but on the other hand I am convinced that it is impossible to give a general abstract proof of the impossibility of hidden parameters or of some other kind of individual interpretation. I think it is only possible to discuss that for a very concrete example, and I do not know any example which I consider satisfactory. If some day somebody does construct such a model, I may be converted, but I do not expect that it will occur.

So two standpoints confront each other: the one is that within the approximation of non-relativistic quantum mechanics the statistical interpretation is the maximal consistent interpretation which one can give, and the other is that even if it might be consistent it is not sufficient and that a more far reaching individual interpretation can or should be found. I propose now to discuss only the first interpretation—the statistical one—as a consistent interpretation, without going into the question whether it is maximal or not, and I shall not discuss any individual interpretation.

In using the statistical interpretation, one should be very careful in one's language. One should not for instance use the hybrid terminology of a system which is in a certain quantum state. That would already suggest an individual interpretation. One should speak instead about a statistical operator which represents an ensemble. The statistical interpretation leaves for example no place for speaking about the quantum state of the universe. In daily life none of us keeps very strictly to the wearisome rules of the statistical interpretation, and it is therefore not surprising that now and then we run against paradoxes. I maintain that such inconsistencies are avoided if we take the trouble to keep very strictly to the statistical interpretation.

The second point has also in a certain way already been discussed by Prosperi. I am afraid I have to repeat some of it, though in a slightly different way. In the first place the statistical operator $\hat{k}$ changes in time, according to the Schrödinger equation. This change can be expressed by

$$\hat{k}(t_2) = \hat{U}(t_2, t_1)\hat{k}(t_1)\,\hat{U}^{-1}(t_2, t_1),$$

where $\hat{U}$ is the unitary evolution operator. The other way in which $\hat{k}$ may change is as the result of a measurement. For the sake of simplicity I consider measurements of the first kind, which formally distinguishes between a complete set of subspaces of the Hilbert space of the statistical operators $\hat{k}$. These subspaces are represented by their projection operators $\hat{P}_m$. Only for a maximal measurement are they one-dimensional, but we allow them to be more- or perhaps many-dimensional as well.

If I make a measurement, then the first step (I) is that the initial statistical operator $\hat{k}$, which may be written

$$\hat{k} = S_{mn}\,\hat{P}_m\hat{k}\,\hat{P}_n$$

changes according to the non-unitary transformation

$$S_{mn}\,\hat{P}_m\hat{k}\,\hat{P}_n \xrightarrow{\text{I}} S_m\,\hat{P}_m\hat{k}\hat{P}_m.$$

The same has been expressed in another notation by Prosperi, and it just means that the coherence between the experimentally distinguished subspaces is destroyed by the measurements. To cut out all kinds of nonsense about quantum theory of living observers let the measurement be automatized, so that for every sample the result is recorded.

The second step (II) is that I read the records. If in the second step I make a selection of those records showing one particular 'm', I select from the ensemble a subensemble as represented by the step (II)

$$S_m \, \hat{P}_m \hat{k} \hat{P}_m \xrightarrow{\text{II}} \hat{P}_m \hat{k} \hat{P}_m / \text{tr}\,(\hat{k}\hat{P}_m).$$

Now the first step (I) is rather simple to explain; the second one (II), on the other hand, is quite difficult.

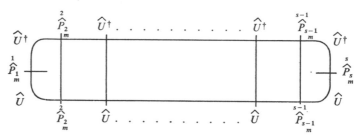

Fig. 2. Relative probability

If I make just one measurement of the whole ensemble, it does not give me much information about the dynamics of the system—mainly about the kinematics. If I want more information about the dynamics, I have to make more measurements at subsequent times. So let me make a series of measurements, which can give answers of quite different types. In the scheme of fig. 2 the various $\hat{P}_m$ denote the projection operators of the various subsequent measurements and $\hat{U}$ denotes the evolution operator for the time intervals between these measurements. The closed line indicates the trace of the product of the operators in the given order. When normalized, this expression gives the statistical frequency of the various answers which you can get from all the readings of all the recordings of the ensemble. As long as the recordings of a measurement, although already made, have not yet been read, we have to take the sum S_m over the corresponding m's.

Notice that the statistical operator does not enter at all. If I want to introduce a statistical operator I may, for instance, read all results recorded before a certain time. From these records I can form a statistical operator which is just the storage of all the information which I have for that time from the earlier measurements. This is not very difficult to explain, but I leave it for the moment.

Alternatively I can read just the results recorded after that time. From these records I can form a different statistical operator—a kind of advanced statistical operator instead of the retarded one—which now stores all the information from the later measurements. From the retarded statistical operator I can calculate the probability of the later results, and from the advanced statistical operator the probability of the earlier results. In both cases I have a statistical operator which represents one or another selection of subensembles and which makes it possible to calculate everything of the statistical probability of the later or of the earlier measurement results. So for a certain time I can assign quite different statistical operators because I make different selections from my ensemble. That is also an argument why the statistical operator is not something which can be ascribed to a single individual system, but only represents some ensemble which is selected with respect to a certain kind of information.

I can do it in still another way: I can just read all the records of some of the measurements and not yet of the other measurements in between. I can calculate the probability of these other measurement results perfectly well with the expression of fig. 2, but it is entirely impossible to introduce ordinary statistical operators at this point. The statistical operator fulfils the general function of a field like the electromagnetic field, for instance, which in a simple efficient way stores the relevant information with respect to the corresponding situation. If I express all past or future information in terms of a retarded- or advanced statistical operator, I can calculate from it all probabilities of future or past measurement results which quantum mechanics is able to derive.

The situation is that we record and read the measurement results in a macrophysical way. From the micro-quantum formalism we derive statistical relations between the recorded measurement results. We may express the micro-quantum formalism in terms of a statistical operator which we assign to the ensemble as a function of time (in the usual forward or the unusual backward order of time—it does not matter). Every time I make a measurement on all the samples of my ensemble I get the readings in macro terms and I make a corresponding selection of my ensemble into subensembles. Then I switch over from the macro-description to the micro-description and assign to the subensemble the corresponding statistical operator and follow its evolution in time. At the next measurement I switch back to the macro-description and calculate the statistical probabilities for the various measuring results. And so on. I switch to and fro between the micro-quantum description and the macro-phenomenological description. I have to come back to that.

Now my third point. If my micro-object system is measured, it is observed by means of some other system and we go on in this way and get the famous von Neumann chain (fig. 3). By amplification the link systems in the chain become larger and larger until we come into the macro-range. If we do not take care, we end in difficulties.

We have to remember that I am not discussing individual samples, but only ensembles: space ensembles of identical systems or time ensembles of repetitions with the same system; it does not matter.

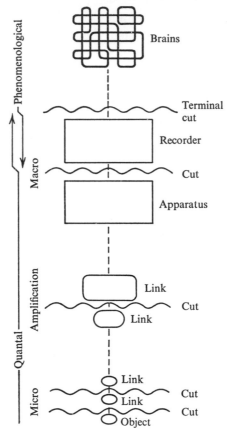

Fig. 3. Measurement chain

It is essential that the difficulties of extending ensembles to brains and their physical description can be avoided. In the von Neumann chain we can make a cut at any place you like, so that we have on the one side the object system and on the other side the observer. Mathematically it does not matter where you make the cut, but physically in a certain way it does matter very much, because we can make a terminal cut in the macro-physical part of the chain (fig. 3).

We must be careful of one thing: in the beginning of the chain we have only micro-systems, and micro-systems are very vulnerable. If the coupling between two successive micro-link systems is not perfectly matched or if there is some external disturbance, the information from the object system will be partially or even entirely lost. Now I want to safeguard the information against destruction by a later observation or an external disturbance. I can do this in a sufficiently advanced part of the chain by an irreversible recording. In order to record in an irreversible way, I have to perform it in a macro-system, because irreversibility is a macro-concept. This invulnerability by means of irreversible processes is, I think, an even more important feature of the macro-character of the measuring and recording systems than the amplification process in the chain. I may record, for instance, by punching holes. Other people like to do it by killing cats.

On the one side of the measuring chain I have the micro-quantum description and on the other side I have the macro-phenomenological description. At the cut I get the switching over between the two descriptions. I think very fundamental problems come in here. One problem we may consider is: we have quantum mechanics, which is initially intended for micro-systems with few degrees of freedom; can we also apply it to the macro-system of the measuring and recording apparatus? If we believe that quantum mechanics is also applicable to large systems with very many degrees of freedom, then we might, in principle, although perhaps not in practice, calculate the statistical operator of this system: the statistical operator, which represents the ensemble of systems we have here.

But even if I were clever enough to calculate such a statistical operator, there is just no connection with the macro-phenomenological description from the other side. If I go through the measuring chain (fig. 3) from the object system as far as the recording system and then forget all systems except the last one, I can speak about this recording system in two different ways: in two different languages. The one is the micro-quantum language and the other is the phenomenological macro-classical language. These two languages are *complementary* in a certain way. The problem is now to find the *correspondence* between these two complementary languages.

After I have derived—maybe only formally and very schematically —the statistical operator for the whole chain in which all successive links have been coupled for some time, I leave out all intermediate links, which now have done their service. Also for the recording instrument I need only very few of the very many degrees of freedom after the recording—the few degrees of freedom which serve the coding of the measuring result. Depending on the context they will—as in the case of punch holes in my tape—be described in macro terms. There remain very many other degrees of freedom, which I do not use any more explicitly, but which I still need in order to take up the disturbances which come from outside or from the observation of the recorded code. If I observe the punch holes, there will be changes in the microdescription, but they are irrelevant for the macro appearance of the holes. So I cannot do without these other degrees of freedom and the intermediate links, but I cannot use them any more. I cannot get further relevant information from them and I cannot control them in an effective way; in order to proceed I have got to ignore them. This means that I have to take the trace of all these degrees of freedom which I do not use—from now on they represent a kind of lost or hidden parameter. A simple consequence is the step (I)

$$S_{mn}\hat{P}_m\hat{k}\hat{P}_n \xrightarrow{\text{I}} S_m\,\hat{P}_m\hat{k}\hat{P}_m.$$

The coherence between different subspaces m and n is lost, because this particular correlation has become extended over all cooperating degrees of freedom. The difficult problem is the remaining correlation between the various Hilbert subspaces $\hat{P}_m$ of the micro-object system and the macro-code in the recording system. This correlation is the purpose of designing a measurement and makes it possible to select in view of the recorded measuring results the corresponding subensembles of the object system as represented by step (II)

$$S_m\,\hat{P}_m\hat{k}\hat{P}_m \xrightarrow{\text{II}} \hat{P}_m\hat{k}\hat{P}_m/\mathrm{tr}\,(\hat{k}\hat{P}_m).$$

I think that up till now I have been largely in agreement with Prosperi's view, but now possibly I come more on Bub's side. I see actually two problems here, which are very closely related and which in my opinion have not yet been solved in a satisfactory way. First the relation between the micro- and macro-observables. This question (i) is how can we relate our punch holes and other phenomenological macro-observables to our formal micro-quantum description with statistical operators $\hat{k}$ and evolution operators $\hat{U}$ and projection operators $\hat{P}_m$? The second question (ii) is a fundamental problem of statistical mechanics: how can we derive from our reversible microquantum description the irreversible macro-equations of hydrodynamics or of elastic bodies or of thermodynamics and so on. The solution of each of these two problems actually requires the answer to the other: in order to solve the second problem I have to know what are my macro-observables, and in order to solve the first problem I have to know how these macro-observables come about in the irreversible recordings. But the situation that they depend on each other does not mean that they are insoluble, it only means that they have to be solved together, perhaps in successive steps.

I think very important work has already been done on the second problem but perhaps not so much on the first problem. Part of this work has been done in Milan by our Italian physicists. In other places work has been done in similar or different ways. I mention, for instance, the approach in terms of infinite systems, with the help of C^* algebra for example. I do have hopes that it will be possible to solve these problems together. For the moment I see them as the main unsolved fundamental problems of quantum theory and in particular of the quantum measuring process.

My last point, number four, is the question about the equivalence or not of the micro-fine-grained statistical operator $\hat{k}$ of a macro-system and the macro-coarse-grained statistical operator $\hat{K}$—which is the W of Prosperi. I first have a comment on the disagreement between Prosperi and Bub. In agreement with what I have said about the use of the statistical operator—that it only applies to an ensemble —I would refuse to talk about the micro-state of a macro-system. I only see a possibility of speaking about a statistical ensemble—of macro-systems in this case. I see the problem in this way: if in some way or other—for instance by a measurement—we have obtained a statistical ensemble of macro-systems represented by a fine grained statistical operator $\hat{k}$, we can neither by information from measurements nor by further control, distinguish it from a statistical ensemble represented by a coarse-grained statistical operator $\hat{K}$, which is obtained from the fine-grained one by some suitable smearing out procedure. If we adopt the attitude of the statistical interpretation, that statistical operators do not describe any kind of states of individual systems, but only represent statistical ensembles selected with respect to some available information or other, then there appears no fundamental difficulty with the effective equivalence of the fine-grained and coarse-grained statistical operators. As soon as it is no longer possible to distinguish between the two with regard to further information and control, it appears not only admissible, but even advantageous, to make the description manageable in replacing the fine-grained operators by coarse-grained ones. It is one of the problems which have also been mentioned in Prosperi's paper and it is narrowly connected with the problems (i) and (ii) which I mentioned earlier in point three. The problem is not only to find the relationship between the fine-grained $\hat{k}$ of the micro-quantum description and the coarse-grained $\hat{K}$ which gets me close to the macro-classical description and which is related to problem (i). I may decide to smear out the fine-grained $\hat{k}$ into a coarse-grained $\hat{K}$ or not at a certain time and then wait some time and decide to smear out again or not and so on. The problem is whether I can in any way distinguish the various descriptions and can derive irreversibility of a Markovian or non-Markovian type (for example, in the form of some master equation). This problem is related to problem (ii).

To end with an example, let us just take a Stern–Gerlach measurement recorded by punch holes on a tape: for spin up I have a hole on the left side, for spin down on the right side. Then the problem of Schrödinger's cat may be formulated in a purely physical (besides milder) version that as long as a punch hole has not yet been observed, we would have a linear superposition of two states with either a hole on the left or a hole on the right. Whereas if you use the statistical interpretation rigorously, you can only speak of a statistical ensemble of samples with either a hole on the left or a hole on the right. The coherence correlation between the two is spread out over intermediate links in the measuring chain and all degrees of freedom of the measuring and recording systems and you cannot get hold of them and design any experiment which shows some kind of interference effects of the left and right holes.

My example is still too schematic. Indeed I think that a significant advance in this field can only be achieved by studying in great detail very concrete examples of physical systems in which measurement is involved. You have to know precisely what you in fact observe. The reason why this has not yet been done in a satisfactory way is that even in the simplest cases we get involved in extremely complicated problems and we have to develop means to solve the problems (i) and (ii).

Addendum on complementarity. At some place in the von Neumann measurement chain which connects the micro-object system and the macro-observer we have to make a cut where we switch over between the micro-quantum formalism on one side and the macro-classical description on the other side. The form in which complementarity comes in, depends on the place where we choose to make the cut.

The simplest choice is to make the cut immediately after the micro-object system and not to analyse any micro-quantum process in the measuring instrument at all. A relatively primitive form of complementarity is that between non-commensurable observables, for example coordinate and momentum, or between the corresponding particle- and wave-pictures. From the point of view of the more complete quantum formalism, the particle-picture is a fair approximation if in the coordinate representation the non-diagonal elements may be neglected and the wave-picture is a fair approximation if in the momentum representation the non-diagonal elements may be neglected. These two conditions cannot occur simultaneously.

In the quantum formalism there appears a kind of complementarity between the unitary transformation $\hat{k}(t_1) \rightarrow \hat{k}(t_2)$ under the influence of an external field and the non-unitary transformation in step (I) under the influence of a measuring instrument (which therefore cannot be represented by an external field). We switch over from the micro-quantum formalism to the complementary macro-phenomenological description by assigning to the measurement result 'm' the statistical probability $tr(\hat{k}\hat{P}_m)$. We switch over from the macro-phenomenological description to the complementary micro-quantum formalism by, after selecting the subensemble corresponding to the reading result 'm', assigning to it the statistical operator obtained in step (II).

If we shift the cut as far as after the macro-recording system, the non-unitary transformation in step (II) originates from the unitary transformation for the interacting systems of the chain by ignoring link systems or degrees of freedom after they have done their duty. The seeming contradiction between the two kinds of transformation has then been solved. Complementarity now comes in at the level of switching over between the micro-quantum description and the macro-classical description of the macro-recorder.

The macro-classical description is adequate to deal with the crude macro-phenomenological aspects of the recording and its observation on the classical side of the cut. But it becomes highly inadequate if we try to extend it as far as the micro-part of the chain and the micro-object system. Then it becomes indispensable to switch over to the refined micro-quantum formalism. On the other hand this quantum formalism is unsuitable for a direct description of the macro-phenomenological aspects on the observers side of the cut. Every time this side comes into consideration it becomes indispensable to switch over to the crude macro-classical description. There is neither a logical contradiction, nor a vicious circle in the alternating use of the complementary classical and quantum descriptions.

I should expect that the question of complementarity would be at least further elucidated as soon as the problems (i) and (ii) have been satisfactorily solved. However, I should not expect that this complementarity would ever really be explained away: I should expect that it would once more be shifted to another level and reappear there in another form.

Almost all our language, concepts and ideas, and habits have been formed under prescientific macro-conditions of daily life and scientific macro-conditions of classical physics. It appears that the crude macro-classical description can adequately be accompanied by a realistic picture. In fact the hypothesis of an objective physical reality seems the best verified hypothesis we know and it is a basic principle of the macro-classical description. There is no guarantee that it is also adequate under the unfamiliar complementary conditions, where in order to relate macro-classical parts of the description of quantal observations, it is indispensable to switch in between these parts the micro-quantum formalism. I do not expect that a realistic picture which might adequately accompany the micro-quantum formalism ever will be found. Anyhow, the statistical interpretation does not require such a micro-picture.

MACROSCOPIC PHYSICS, QUANTUM MECHANICS AND QUANTUM THEORY OF MEASUREMENT

G. M. PROSPERI

In macroscopic physics the behaviour of a macroscopic body is described in terms of a limited number of variables to which are ascribed at any time well defined values, and which evolve deterministically according to certain differential equations. Such variables are called *macroscopic variables* in this paper, and every set of values for them is said to specify a *macroscopic state*.

In quantum mechanics, which applies to microscopic objects (elementary particles, atoms, molecules), one has to work instead entirely in terms of the ideas of observation and probability. The set of abstract mathematical rules of which quantum mechanics consists allows only statistical predictions, and the appearance of the so called *interference terms* in the expression of the transition probabilities prevents one from ascribing to a certain quantity a value independent of an actual observation of it.

However, since a macroscopic body is made up of atoms and molecules, the macroscopic and the quantum descriptions should not be independent and we are faced with the problem of deriving the laws of macroscopic physics from the laws that control the behaviour of the elementary components of a large body. In its most general aspect this is the problem of statistical mechanics. This problem cannot be considered completely solved today, although there are a number of relevant results in explaining the macroscopic properties of certain specific systems in equilibrium and in non-equilibrium and in evaluating certain quantities which macroscopic theory has to take from experience (thermodynamic function, transport coefficients, etc.).[1]

The problem is of fundamental importance for the interpretation of quantum mechanics itself. In fact, while the macroscopic variables for a macroscopic body describe something which is connected with our common experience in a simple way, the microscopic objects can be observed only by the modifications that they produce on macroscopic bodies. In quantum mechanics an 'observable quantity', A, is described by means of an abstract mathematical tool, a *self-adjoint operator*, acting on an appropriate Hilbert space. An apparatus to measure the quantity A is a macroscopic system, the macroscopic state of which is modified by the interaction with the object observed in one or other defined way according to the state vector of the object coinciding with one or other eigenvector of A. Consequently an analysis of the process of measurement is necessary if we want to establish which self-adjoint operator has to be associated with a certain definite kind of measurement.

Such an analysis is always present, at least implicitly, every time the abstract formalism is used in connection with actual experiments.

In deriving macroscopic physics from quantum mechanics we are faced with two separate problems.

(1) To construct the self-adjoint operators that describe the macroscopic quantities in the language of quantum mechanics.

(2) To show that the dispersion for the macroscopic quantities can be made negligible on the macroscopic scale and that the practically well defined value for such quantities obeys the correct equations of macroscopic physics.

The solution of problem 1 is practically determined by the correspondence principle. The most common macroscopic variables for a homogeneous system in equilibrium conditions are total energy, total momentum, electric charge, chemical composition, etc. and for a system in non-equilibrium, the same quantities for small parts of it. We know very well how to construct operators that represent such quantities in quantum mechanics. In this analytical representation there is, however, a large freedom due to the limited accuracy of the macroscopic observations, and the choice of the most convenient formalism in the non-equilibrium case is not a trivial problem—particularly if we want to describe the system macroscopically as continuous rather than as an ensemble of a finite number of parts.

We shall not bother further in this paper with problem 1; we want instead to discuss briefly some general aspects of problem 2 in connection with the problem of measurement in quantum theory.

The most striking fact concerning problem 2 is the possibility of a closed description of the behaviour of the system in terms of macroscopic quantities alone; i.e. the fact that a knowledge of the values of the macroscopic quantities at a certain time is sufficient to calculate the values of the same quantities at any subsequent time.

Let us now symbolically denote by M the set of operators that represent the macroscopic quantities for a large body and write the eigenvalues equation

$$M |\Omega_{\nu i}\rangle = m_\nu |\Omega_{\nu i}\rangle \quad i = 1, 2, ..., s_\nu \tag{1}$$

where i is the degeneration index. Let us assume that we have performed an observation of M at the time $t = 0$ and let us represent the information that we have obtained by a distribution of probability u_ν^0 rather than by a single value m_ν; this fits better with the fact that macroscopically we must look at M as a set of quantities assuming continuous rather than discrete values.

According to the postulates of quantum mechanics if we do not have any other information about the system we must represent the situation immediately after such observation by the statistical operator

$$W(0) = \Sigma_\nu u_\nu^0 \frac{1}{s_\nu} P_\nu, \tag{2}$$

where

$$P_\nu = \sum_{j=1}^{s_\nu} |\Omega_{\nu j}\rangle \langle \Omega_{\nu j}|$$

is the projector on the νth eigenmanifold of M.

The statistical operator at the time t is then

$$W(t) = e^{-iHt} W(0) e^{iHt} \tag{3}$$

and the probability of finding for M the value m

$$u_\nu(t) = \mathrm{Tr}\,(P_\nu W(t)). \tag{4}$$

By a comparison with equation (2) we are brought now to consider the new statistical operator

$$\tilde{W}(t) = \Sigma_\nu u_\nu(t) \frac{1}{s_\nu} P_\nu \tag{5}$$

and for an arbitrary function of M we have obviously

$$\langle f(M) \rangle_t = \mathrm{Tr}\,(f(M)\,W(t)) = \mathrm{Tr}\,(f(M)\,\tilde{W}(t)). \tag{6}$$

According to a terminology introduced by Jauch[2] we may say that $W(t)$ and $\tilde{W}(t)$ are *macroscopically equivalent* and write

$$W(t) \approx \tilde{W}(t). \tag{7}$$

We note that equation (6) is a completely general and in some sense trivial statement in which no specific property of the observables M is implied. In particular we want to stress that, if $t_0 > 0$ is an arbitrary fixed time, equation (6) states that the two statistical operators $W(t_0)$ and $\tilde{W}(t_0)$ give the same statistical predictions as far as the observable M is concerned. Their time evolutions at a subsequent time t on the other hand give, *a priori*, completely different predictions; the expressions for $u_\nu(t)$ as evaluated from $W(t_0)$ and $\tilde{W}(t_0)$ differ by the already mentioned interference terms.

The possibility of a description of the behaviour of a large system in terms of macroscopic quantities alone suggests however that in such a case the equivalence between $W(t_0)$ and $\tilde{W}(t_0)$ is 'conserved' during time evolution; i.e. that for any fixed $t_0 < t$ the equation

$$e^{-iH(t-t_0)}\, \tilde{W}(t_0)\, e^{iH(t-t_0)} \approx W(t) \tag{8}$$

holds to a good approximation or equivalently that the interference terms are negligible.

Equation (8) expresses essentially the fact that the distribution of probability $u_\nu(t)$ can be expressed in terms of the distribution of probability $u_\nu(t_0)$ and the difference $t - t_0$ alone, while any other information that we may have on the system is irrelevant and the initial time $t = 0$ at which equation (2) is assumed does not play any privileged role.

According to the terminology of the probability calculus such a process is called a *Markoff process*.[3] It can be easily shown that for such a process a differential equation of the form

$$\frac{du_\nu(t)}{dt} = \Sigma_{\nu'} Q_{\nu\nu'} u_{\nu'}(t) \tag{9}$$

has to hold. Such an equation is called the *master equation* or *Kolmogorov equation*, and in the context of statistical mechanics was first proposed by Pauli in 1928.[4]

In the approximation in which M is treated as a continuous variable, equation (9) can be also written

$$\frac{\partial u(m, t)}{\partial t} = \int dm'\, (Q(m, m')\, u(m', t) - Q(m', m)\, u(m, t)). \tag{9'}$$

From equation (9'), under the assumption that the deviation $\langle (M - \langle M \rangle_t)^2 \rangle$ of M from its expectation value $\langle M \rangle_t$ remains small as t increases, the following new equations can be deduced[5]

$$\frac{d\langle M \rangle_t}{dt} = \alpha(\langle M \rangle_t), \tag{10}$$

$$\frac{d\langle (M - \langle M \rangle_t)^2 \rangle}{dt} = \beta(\langle M \rangle_t) + 2\, \frac{\partial \alpha(\langle M \rangle_t)}{\partial m} \langle (M - \langle M \rangle_t)^2 \rangle \tag{10'}$$

where
$$\alpha(m) = \int dm' Q(m', m)\, (m' - m), \tag{11}$$

$$\beta(m) = \int dm' Q(m', m)\, (m' - m)^2. \tag{11'}$$

Equation (10) is of the form occurring in macroscopic physics, while equation (10') allows one to check that the dispersion actually remains small during the time evolution. Our problem 2 is consequently completely solved if we can prove equation (9) for our quantity M. Equation (9) however can only be an approximate equation. An exact equation of somewhat similar form for $u_\nu(t)$ can be derived from the Liouville–von Neumann equation (or from equation (3)) and from the initial condition expressed in equation (2); it has the form[6]

$$\frac{du_\nu(t)}{dt} = \Sigma_{\nu'} \int_0^t dt'\, K_{\nu\nu'}(t')\, u_{\nu'}(t - t'), \tag{12}$$

where the nucleus $K_{\nu\nu'}(t)$ is expressed simply in terms of the Hamiltonian of the system.

Equation (12) is of a quite different nature from (9). The right-hand side contains $u_{\nu'}(t')$ in the entire interval $0 \leqslant t' \leqslant t$ and the $u_\nu(t)$'s cannot any longer be expressed in terms of the $u_\nu(t_0)$'s and $t - t_0$ alone.

The time $t = 0$ has a very special position in these equations. This is not surprising since equation (12) in contrast with (9) is purely formal and holds for any observable and system. For a large system however the nucleus $K_{\nu\nu'}(t)$ has a special structure. As a consequence of the practically continuous nature of the energy spectrum it vanishes as t becomes large compared with a certain characteristic time t_m. The time t_m usually equals the characteristic time of some elementary microscopic process; for a dilute gas for instance t_m is of the order of the duration of a collision between two molecules. Consequently one expects t_m to be very small in general compared with the characteristic time t_M for a macroscopic modification in the body to occur.†

$$t_m \ll t_M. \tag{13}$$

At first sight one would also expect the characteristic time t'_M for a significant change in $u_\nu(t)$ to be of the same order of magnitude as t_M. If it were so, it would be possible to replace $u_{\nu'}(t - t')$ in equation (12) by $u_{\nu'}(t)$ and the upper limit in the integral by $+\infty$. Equation (12) would then become identical with (9) with

$$Q_{\nu\nu'} = \int_0^\infty dt'\, K_{\nu\nu'}(t'). \tag{14}$$

Actually the situation is not so simple. Taking into account equations (14) and (10) we have $t'_M \sim 1/Q_{\nu\nu}$ and $t_M \sim 1/\alpha'(m)$ and these quantities can be evaluated in significant examples. The results suggest that the relation $t'_M \sim t_M$ holds only in very particular cases;

† A counter example is given by the relaxation effect of the nuclear magnetization[10] which is not however a really macroscopic quantity in the strict sense we are using here.

precisely, when the observables M specify the state of a small part of the system in weak interaction with the remaining part. If instead M is a collective variable for the entire system one finds $t'_M \sim t_M/N$, where N is the number of elementary components of the system. In the thermodynamic limit then $(N \to \infty,\ N/V \to \text{const.})$ t_M remains finite, while t'_M vanishes.

The difference in order of magnitude between t'_M and t_M is due to the particular structure of $Q(m, m')$. In the examples that have been studied, this quantity is different from zero only for m' very close to m and consequently strong cancellations occur in equation (11). It can be seen from (10') that the characteristic time for $\langle (M - \langle M \rangle_t)^2 \rangle$ is of the same order as t_M. The same situation occurs in general for expressions involving low order momenta. As the order of the considered momentum increases however the characteristic time becomes progressively closer and closer to t'_M.

If we now try to obtain directly from equation (12) equations involving $\langle M \rangle_t$ and $\langle (M - \langle M \rangle_t)^2 \rangle$, we obtain instead of (10) and (10') certain *integro-differential equations* the nuclei of which decay in the same time t_m as $K_{\nu\nu'}(t)$. Consequently we can repeat for such equations the argument given above for the derivation of equation (9) from (12) when $t'_M \sim t_M$ and we obtain equations (10) and (10') with $\alpha(m)$ and $\beta(m)$ given by (11), (11') and (14).

Then, even if equation (9) does not give in general a good approximation for $u_\nu(t)$, nevertheless, if equation (13) holds, (10), (10') and the other analogous ones give good approximations for $\langle M \rangle_t$, $\langle (M - \langle M \rangle_t)^2 \rangle$, etc., or, what amounts to the same, equation (9) gives correctly the lowest order momentum even if it is not a good approximation for the distribution of probability $u_\nu(t)$ in itself. [7]

The above results can also be expressed by saying that the two sides of equation (8) give the same expectation value for a polynomial in M of low degree, but not for every conceivable function of M. We shall say that two statistical operators in such a relationship are *weakly equivalent* and shall write

$$e^{-iH(t-t_0)}\, \tilde{W}(t_0)\, e^{iH(t-t_0)} \overset{\text{weak}}{\approx} W(t), \qquad (15)$$

which has the same physical content as equation (8). We shall also say that equation (9) holds in the *weak sense*.

According to our discussion, the solution of problem 2 is reduced to proving equation (13). Such a proof obviously cannot be given in general, but requires a detailed study of specific examples. In particular a consideration of the specific properties of M is essential.

Let us now briefly look at the role of equation (15) in the quantum theory of measurement. An apparatus for the observation of a microscopic object can be described as a macroscopic body which possesses a second macroscopic constant of the motion J beside the energy E. It is assumed that:

(1) the interaction between the apparatus and the object is effective for a time τ so short that no appreciable modification of the quantity A which we are measuring or of the macroscopic state of the apparatus can occur in the absence of the interaction during that time;

(2) before the interaction the apparatus is in the equilibrium state corresponding to certain given values E_0 and J_0 of the macroscopic energy and of the constant J;

(3) as a consequence of the interaction J is changed to new values J_k in one to one correspondence with the eigenspaces of A;

(4) ultimately the apparatus reaches the equilibrium state corresponding to the new values of J, E remaining unchanged.[8]

In the following we shall denote by I and II the object and the apparatus respectively, and use such labels to distinguish the quantities referring to the one or the other of such systems. We shall write them

$$H = H^{\mathrm{I}} + H^{\mathrm{II}} + H^{\mathrm{int}} \tag{16}$$

for the Hamiltonian of the total system,

$$A^{\mathrm{I}} |\phi_h^{\mathrm{I}}\rangle = a_h |\phi_h^{\mathrm{I}}\rangle \tag{17}$$

for the eigenvalue equation for A and

$$\left. \begin{aligned} J^{\mathrm{II}} |\Omega_{kvj}^{\mathrm{II}}\rangle &= J_k |\Omega_{kvj}^{\mathrm{II}}\rangle \\ M^{\mathrm{II}} |\Omega_{kvj}^{\mathrm{II}}\rangle &= m_v |\Omega_{kvj}^{\mathrm{II}}\rangle \end{aligned} \right\} \tag{18}$$

in substitution for the eigenvalue (6). Here we have explicitly introduced a new index referring to the values of J (the explicit introduction of an index referring to the value of the macroscopic energy is immaterial since this quantity does not change during the process).

Let us now conventionally denote by $t = 0$ the time at which the observation is performed, i.e. the time at which the interaction between the object and the apparatus begins to become effective, and let us write

$$|\psi_0^{\mathrm{I}}\rangle = \Sigma_h c_h |\phi_h^{\mathrm{I}}\rangle \tag{19}$$

as the state vector of the object at $t = 0$. The statistical operator for the total system I + II at the same time can then be written

$$W(0) = \sum_{hh'} c_h c_{h'}^* |\phi_h^{\mathrm{I}}\rangle \langle \phi_h^{\mathrm{I}}| \otimes \frac{1}{s_{0e_0}} \sum_{j=1}^{s_v} |\Omega_{0e_0 j}^{\mathrm{II}}\rangle \langle \Omega_{0e_0 j}^{\mathrm{II}}| \tag{20}$$

where by e_0 we have denoted the value of ν corresponding to the equilibrium state for $J^{II} = J_0$.

According to assumption (3) we have

$$e^{-iH\tau}|\phi_h^I\rangle\otimes|\Omega_{0e_0j}^{II}\rangle = |\phi_h^I\rangle\otimes|\Phi_{hj}^{II}\rangle = |\phi_h^I\rangle\otimes\sum_{\nu l}\alpha_{hj}^{\nu l}|\Omega_{h\nu l}^{II}\rangle \qquad (21)$$

and consequently

$$W(\tau) = \sum_{hh'}c_h c_{h'}^*|\phi_h^I\rangle\langle\phi_{h'}^I|\otimes\frac{1}{s_{0e_0}}\sum_j|\Phi_{hj}^{II}\rangle\langle\Phi_{hj}^{II}| \qquad (22)$$

and $$W(t) = \sum_{hh'}c_h c_{h'}^*\exp(-iH^It)|\phi_h^I\rangle\langle\phi_{h'}^I|\exp(iH^It)$$

$$\otimes\frac{1}{s_{0e_0}}\sum_j\exp(-iH^{II}t)|\Phi_{hj}^{II}\rangle\langle\Phi_{h'j}^{II}|\exp(iH^{II}t), \qquad (23)$$

where the free time evolution during the interval τ has been consistently neglected.

We now introduce a notion of equivalence for two statistical operators describing the total system $I + II$. We shall call W_1 and W_2 *equivalent* if they give the same expectation value for operators of the form $B^I\otimes f(J^{II}, M^{II})$, where B^I is an arbitrary operator for I and $f(J^{II}, M^{II})$ is an arbitrary function of J^{II} and M^{II}: if the same is true under the assumption that $f(J^{II}, M^{II})$ is a polynomial of low degree in M^{II} we shall call W_1 and W_2 *weakly equivalent*.

Let us then consider the statistical operator

$$\tilde{W}(t) = \sum_h|c_h|^2[\exp(-iH^It)|\phi_h^I\rangle\langle\phi_h^I|\exp(iH^It)\otimes\sum_\nu u_{h\nu}(t)P_{h\nu}^{II}], \quad (24)$$

where $$u_{h\nu}(t) = \frac{1}{s_{0e_0}}\sum_{jl}|\langle\Omega_{h\nu l}^{II}|\exp(-iH^{II}t)|\Phi_{hj}^{II}\rangle|^2$$

we have obviously $\tilde{W}(t) \approx W(t)$. By arguments of the type used above it can be shown furthermore that even if $W(\tau)$ is not, as far as II is concerned, strictly of the form (2), equation (15) holds.[9]

If, in considering II, we are interested only in the macroscopic state, then the interference terms can be neglected, the statistical operator $W(t)$ can be replaced in every effect by $\tilde{W}(t)$ and $u_{h\nu}(t)$ can be replaced by the solution of an equation of the form (9). This result is consistent with all reasonable requirements for a theory of the measurement process. In particular it signifies that, although the interaction between the object and the apparatus is an essentially microscopic process, yet as soon as the interaction is no longer effective the apparatus can be described by macroscopic physics and the results expressed in terms of our common experience.

REFERENCES

(1) Landau, L. P. and Lifschitz, E. *Statistical Physics* (London: Pergamon Press, 1959).

Huang, K. *Statistical Mechanics* (New York: Wiley, 1967).

Bak, Thor A. (ed.) *Statistical Mechanics, Foundation and Applications* (New York: Benjamin, 1967).

Zwanzig, R. *A. Rev. phys. Chem.* (1965) **16**, 67.

Fisher, M. E. *Rep. Progr. Phys.* (1967) **30**, II, 615.

(2) Jauch, J. M. *Helv. phys. Acta* (1964) **37**, 193.

(3) Feller, W. *An introduction to probability theory and its applications* (New York: Wiley, 1950).

(4) Pauli, W. *Sommerfeld Festschrift* (Leipzig, 1928).

Van Kampen, N. G. *Physica* (1957) **23**, 707 and 816. *Physica* (1959) **25**, 1294. *Fundamental Problems in Statistical Mechanics* (editor E. G. D. Cohen; Amsterdam: North Holland, 1962).

Van Hove, L. *Physica* (1955) **21**, 517.

Prigogine, I. *Non Equilibrium Statistical Mechanics* (New York: Interscience, 1962).

(5) Van Kampen, N. G. See reference (4).

De Groot, S. R. and Mazur, P. *Non Equilibrium Thermodynamics* (Amsterdam: North Holland, 1962).

(6) Zwanzig, R. *J. chem. phys.* (1960) **33**, 1338. *Lect. theor. Phys.* (Boulder, 1960) **3**, 106. *Phys. Rev.* (1961) **124**, 983.

Montroll, E. W. *Lect. theor. Phys.* (Boulder, 1960) **3**, 221. *Fundamental Problems in Statistical Mechanics* (editor E. G. D. Cohen; Amsterdam: North Holland, 1962).

Ludwig, G. *Z. Phys.* (1963) **173**, 232.

Emch, G. *Helv. phys. Acta* (1964) **37**, 532.

Cf. also:

Van Hove, L. *Physica* (1957) **23**, 441.

Prigogine, L. and Resibois, P. *Physica* (1961) **27**, 629.

Resibois, P. *Physica* (1961) **27**, 541. *Physics of Many Particles Systems* (editor E. Meeron; New York: Gordon and Breach, 1966) 495.

(7) Van Hove, L.; Montroll, E. W.; Prigogine, I. and Resibois, P.; Resibois, P.; Ludwig, G. See reference (6).

Lanz, L., Ramella, G. and Scotti, A. *Statistical Mechanics* (editor Thor A. Bak; New York: Benjamin, 1967) 32.

Lanz, L. and Ramella, G. Preprint IFUM 078/FT (Milano, July 1968).

(8) Ludwig, G. *Z. Phys.* (1953) **135**, 483. *Die Grundlagen der Quantenmechanik* (Berlin: Springer–Verlag, 1954), chap. 5. *Phys. Bl.* (1955) **11**, 489.

Daneri, A., Loinger, A. and Prosperi, G. M. *Nucl. Phys.* (1962) **33**, 297. *Nuovo Cimento* (1966) **44**, 119.

Prosperi, G. M. *Encyclopedic Dictionary of Physics* (Oxford: Pergamon Press) suppl. vol. 2.

Rosenfeld, L. *Suppl. Progr. Theor. Phys.* (1965) extra n. 222.

Cf. also:

von Neumann, J. *Mathematical Foundations of Quantum Mechanics* (Princeton: Princeton University Press, 1955), chaps. 5 and 6.

Jordan, P. *Phil. Sci.* (1952) **16**, 269.

Wigner, E. P. *Z. Phys.* (1952) **133**, 101. Am. *Journ. Phys.* (1963) **31**, 755.

Jauch, J. M. See reference (2).

Jauch, J. M., Yanase, M. M. and Wigner, E. P. *Nuovo Cimento* (1968) **48**B, 144.

Loinger, A. *Nucl. Phys.* (1968) A **108**, 245.

Rosenfeld, L. *Nucl. Phys.* (1968) A **108**, 241.

D'Espagnat, B. *Conceptions de la physique contemporaine* (Paris, 1965).

Bub, J. *Nuovo Cimento* (1968) **57**B, 503.

(9) Sabbadini, A. Thesis (unpublished).

Lanz, L., Prosperi, G. M. and Sabbadini, A. In preparation.

(10) Philippot, J. and Walgraff, D. *Physica* (1966) **32**, 1283.

COMMENT ON THE
DANERI-LOINGER-PROSPERI
QUANTUM THEORY OF MEASUREMENT

JEFFREY BUB†

The measurement problem of the quantum theory is the problem of providing an explanation for the projection or 'collapse' of the Hilbert space vector—the so-called 'quantum state'—onto a particular member of a certain relevant set of eigenvectors during a measurement process—representing a stochastic change, which *prima facie* is inconsistent with the unitary time transformations of the theory. I call this a 'problem' in the quantum theory, because the process of measurement in the classical theory is subsumed under the general equation of motion of the theory—measurement has no special status in the theory. Hence, a proposed 'solution' to this problem will either provide an argument to the effect that the measurement process can in fact be subsumed under the unitary time transformations of the quantum theory, or provide an interpretation of the quantum theory as a new kind of description which differs fundamentally from the classical description, so that the 'measurement problem' as I have formulated it does not arise. The former approach involves treating the quantum theory as a particular mechanical theory, appropriate to the mechanical systems of the micro-domain, with an ontological significance analogous to classical mechanics. This is the approach of von Neumann. The latter approach is that of Bohr.

I believe that there can be no 'solution' to this 'problem' without recognizing, with Bohr, the essential difference between the kind of description involved in the quantum theory, and the kind of description involved in the classical theory. The Daneri–Loinger–Prosperi theory is presented as a solution to the measurement problem, understood in the sense of von Neumann. If the Daneri–Loinger–Prosperi theory of measurement is acceptable, then von Neumann's basic approach is vindicated, and the quantum theory can 'stand on its own feet'—all the apparent puzzles of the theory as formulated by von Neumann have been solved. I want to show that the Daneri–

† Supported by the National Science Foundation.

Loinger–Prosperi theory is inadequate as a solution to the measurement problem of the quantum theory, and that no refinement or elaboration of this theory can solve the problem.

Prosperi has given an exposition of the main points of the Daneri–Loinger–Prosperi theory. The quantum measurement process is treated as a *statistical mechanical problem*—a problem in the statistical mechanics of the quantum theory. This involves a quantum ergodic theory of macrosystems, which in principle replaces classical mechanics as the fundamental theory of the macrolevel. The measuring instrument is regarded as a macrosystem for which macrostates can be defined in some way within the framework of formal quantum theory. As in classical statistical mechanics, these macrostates are related to probability distributions over 'quantum microstates' of the system (i.e. complete state descriptions in the quantum mechanical sense, by means of which the system is represented by a vector in Hilbert space).

The idea seems to be that the detection of a microsystem by a macro-instrument always involves a process of *amplification*, which can be understood as the irreversible transition of the amplifying apparatus to a condition of stable equilibrium. It is supposed to be in virtue of this irreversible transition that the linear superposition of Hilbert space vectors (representing the quantum microstate of the composite system: microsystem I + macro-instrument II, after the measurement interaction) can be replaced by a statistical distribution of quantum microstates. The aim of the theory is to show that this statistical distribution coincides with the distribution of eigenstates predicted by the algorithm of the quantum theory.

In order to see the inadequacy of this approach to the measurement problem, it is necessary to consider carefully the status of *macrostates* in the context of the theory. Unless it can be shown that the macrostates are 'objective', in the sense that a particular macroscopic characterization of a system is independent of the experimental context and strictly excludes microstates which can be expressed as linear superpositions of Hilbert space vectors associated with different macrostates, the theory fails. The question which Prosperi's discussion raises is essentially this: how does it *follow from the theory* that the macrostates of large bodies are 'objective', so that they can play an analogous role to classical macrostates in the interpretation of the theory? In other words, how does the *theory* guarantee that quantum theoretical macrostates will not exhibit interference effects?

I shall refer to the original paper of Daneri, Loinger, and Prosperi.[1] In the introduction to this paper, they state:

In order that an objective meaning may be attributed to the macrostates of the large bodies, it is of course necessary that—by virtue of the laws of quantum mechanics and of the structure of the macroscopic bodies—states incompatible with the macroscopic observations be actually impossible.

Call this Thesis 1. This simply means that it should follow from the theory that the only possible *microstates* of a *macrosystem* are those microstates which are compatible with macroscopic observations. In other words, the only possible Hilbert space vectors for a macrosystem should be those belonging to the orthogonal manifolds into which the macroscopic observations decompose the Hilbert space. Since linear combinations of vectors from different manifolds would be impossible—physically impossible—the macrostates would then be 'objective' in the sense that they would define a set of mutually exclusive and collectively exhaustive possibilities for the system, independent of any individual observer.

In fact, this thesis is not proved directly. What is actually proved, is the following: that in virtue of the laws of quantum mechanics and of the decomposition of Hilbert space into cells by the assumption of macro-observations, states incompatible with macro-observations are actually *macroscopically unobservable*. Call this Thesis 2. This thesis does not deny the existence of dynamical states of the system which are incompatible with macro-observations—in other words that the Hilbert space vector representing the system might actually be a linear combination of vectors from different manifolds. It asserts only that these states are not observable by a macro-observer using a macro-apparatus. It appears that Daneri, Loinger, and Prosperi have confused these two theses.

In their original paper (p. 315), the following argument is proposed for the impossibility of observations incompatible with the macro-observations:

Let us observe, however, that in this way the body would be dissolved into its constituents; the ergodicity conditions by virtue of which we have been able to exhibit the impossibility of non-macroscopic measurements on a large body are, on the contrary, just a consequence of the interactions which exist between the particles of the macroscopic body and which have the effect of lumping the particles into a single whole. The meaning of our considerations is precisely the following: when the particles of a system do interact so strongly that a body is formed, it is no longer possible to think of observing them independently.

According to Daneri, Loinger, and Prosperi, the physical structure of a large body (i.e. the interactions between the particles) implies the ergodicity conditions, and the ergodicity conditions imply the impossibility of non-macroscopic *measurements* on a large body. So, the impossibility of non-macroscopic *measurements* on a large body is a consequence of the physical structure of large bodies. Macroscopically unobservable states are not *possible physical states* of the *composite system*: macrosystem II + macro-instrument III (macrosystem II is the macro-instrument measuring microsystem I). This implies that macroscopically observable states are the only possible physical states of such systems.

To put this another way: Thesis 1 says that the total set of possible dynamical states (in other words, the total set of quantum mechanical microstates) of a macrosystem can be subdivided into subsets, each of which corresponds to a macrostate of the system; there are no other possible dynamical states of a macrosystem. This is not proved directly. Instead, Daneri, Loinger, and Prosperi attempt to prove Thesis 2, which asserts that all dynamical states not belonging to a subset of microstates corresponding to a macrostate are unobservable under certain conditions. And finally they argue that these states are unobservable *because* they are not possible dynamical states of the combined macrosystem II and macro-instrument III, which leaves only the microstates associated with the set of macrostates as the set of possible microstates for macrosystem II.

What I find unsatisfactory about this argument (as support for the plausibility of Thesis 1) is the necessity to relate the macrostates to interactions between a macrosystem and a macro-instrument, so that the 'objectivity' of the set of macrostates is only established with respect to a certain macro-instrument. The set of macrostates is only a mathematical device, *which is irrelevant to the actual or possible physical states of the system*, unless a macro-observer 'measures' the corresponding observables with a suitable macro-instrument. I do not think it is sufficient in this context to appeal to the structure and function of the first macrosystem as a measuring instrument II of some measured microsystem I, because this is precisely what is left out of formal quantum theory and what Bohr's interpretation attempts to supply with the argument for a fundamental distinction between the measuring apparatus and the objects under investigation. So ultimately, even introducing a second macro-instrument III to establish the 'objectivity' of the macrostates of the first macro-instrument II does not achieve anything, if this system III is also treated formally within the framework of the quantum theory. Because unless the *observer* defines the structure and function of system III as a particular kind of *macro-instrument*, there is no reason why the actual states of system III should be compatible with the set of possible macrostates defined by the macro-observables, and hence there is also no reason why the actual states of system II should be compatible with the set of macrostates. And I take it that the whole point of the Daneri–Loinger–Prosperi theory is to avoid introducing the observer in this *ad hoc* way.

It is clear that one of the motivations for applying the quantum ergodic theory of macrosystems to the measurement process is a dissatisfaction with von Neumann's solution to the measurement problem. Von Neumann demonstrated the consistency of the projection postulate with the unitary time transformations of the theory. If system I is measured by a composite system (II + III), then the projection postulate allows the replacement of an initial statistical operator, representing a 'pure state' for system I, by an appropriate final statistical operator, representing a 'mixed state' for system I. This replacement is consistent—for measurements on system I—with the assumption of a unitary time transformation for the statistical operator associated with the composite system (I + II), and a measurement of this system by system III (i.e. an application of the projection postulate to the transformed statistical operator for system (I + II)). However, this argument does not demonstrate that a statistical operator representing a pure state can be replaced by a statistical operator representing a mixed state after some particular stage—unless this stage involves the 'mind of the observer', which is a ridiculously *ad hoc* and illegitimate assumption in this context. It is because Daneri, Loinger, and Prosperi believe that the final stage of a measurement is a physical process described theoretically by the transition from a statistical operator representing a pure state to a statistical operator representing a mixture, that they attempt to provide this theoretical description by treating the measurement process as an interaction between a microsystem and a macro-instrument, according to their quantum ergodic theory of macrosystems. The pure state is then replaced by a mixture at the first stage in the measurement process which involves a macrosystem.

To sum up: it seems from Prosperi's argument that one can cut von Neumann's infinite regress at the first macro-instrument, because it follows purely on the basis of the quantum theory (supplemented by the Daneri–Loinger–Prosperi *quantum* ergodic theory of macrosystems) that the macrostates of this system are 'objective'. But the macrostates of the first macro-instrument (system II) are claimed to be 'objective' only in the sense that microstates expressible as linear combinations of vectors corresponding to different macrostates are not possible physical states of the composite system: system II + system III (where system III is a particular kind of macro-instrument capable of measuring the macro-observable corresponding to the macrostates of system II). And the assertion that system III is a macro-instrument with this particular structure and function (i.e. that a certain set of macrostates is relevant for system III) means in turn that microstates expressible as linear combinations of vectors corresponding to different macrostates are not possible physical states of a composite system: system III + system IV (where system IV is a particular kind of macro-instrument capable of measuring the macro-observable corresponding to the macrostates of system. III). And so on. This is von Neumann's infinite regress all over again.

I have pointed out that it is not possible to define 'objective' macrostates, analogous to classical macrostates, without adding some new principle to the quantum theory. And this *is* the *measurement problem* of the quantum theory.

REFERENCE

(1) Daneri, A., Loinger, A. and Prosperi, G. M. *Nucl. Phys.* (1962) **33**, 297.

THE PHENOMENOLOGY OF OBSERVATION AND EXPLANATION IN QUANTUM THEORY

J. H. M. WHITEMAN

The formalism of quantum theory is not, in itself, a physical theory, but it becomes one by virtue of an epistemology relating it to certain kinds of perceptual experience as *explanans* to *explanandum*. Such epistemology involves much more than mere operational rules. There is, for instance, the epistemology of Bohr, which can be called 'phenomenological' to the extent that he insisted on the Gestalt character (wholeness) of each experimental set-up and the uniqueness of each occasion for the providing of data: after which one can consider the limitations on the nature of the constructs of any theory proposed as *explanans*.

In contrast, the classical tradition of simply-located objects characterised independently of experiment† was presupposed by Born and von Neumann and imposed on the data with the help of an informal language of 'particles' and 'states'.‡ Hence arise puzzles and paradoxes such as the so-called 'measurement problem' (reduction of the state-vector) and the problem of 'duality'. Bohr's rejection of the classical ontology was not wholly adequate to solve these problems, if only because primitive terms such as 'observation', 'measurement' and 'particle' had not been given phenomenologically clear and unambiguous usages. It is the purpose of this short paper to outline a more thoroughgoing phenomenological analysis, in the hope that the way may then be cleared for a more soundly-based and open consideration of existing formalisms and of possible advances beyond.

Observation: denotation of the term. Observation can be qualitative or quantitative. For the purposes of the present discussion

---

† Von Weizsäcker referred in the discussions during the colloquium to 'the classical ontology to which von Neumann is fully committed'. 'The problem of measurement certainly is not sufficiently described in von Neumann's way, because there is not the confrontation between one object which is the quantum object and another object which is either the measuring device or the human being.'

‡ Cf. F. Seitz's reference[1] to 'the deductive logic of the ancient philosophers, who began with generalizations and tried to fit these to the facts', in contrast to the modern 'drawing of general principles from specific facts'.

I consider a quantitative result essential, while qualitative aspects are for the purpose of recognition, preliminary classification, and part-confirmation. I also take observation as being 'direct', i.e. by the eye, touch or other senses, with or without the interposition of simple optical appliances and with or without the means of a mechanical or photographic record reasonably considered reliable and not open to varying interpretations. The quantitative result is to be obtained by the use of local Euclidean geometry and regular time-measurement, without the introduction of theories or mechanical aids other than well-established methods of correction to ensure the validity of the local Euclidean geometry and regularity of time-measurement.

Observation is to be made on an experiment, and refers primarily to a focal region but also to the total conditions. The term 'experiment' implies that a part of the total conditions can be supposed constant under observation, while other parts vary with or without prior adjustments.

Among 'prior adjustments' may be included the injecting of 'materials' for interaction with the rest of the conditions, the presumed injecting of such materials in the form of radiation, or the setting up of fields of any kind.

I shall admit two extensions of the above strict concept of observation. One is the attaching of a measure of frequency or wavelength to the observation of a spectral line. I take this as sufficiently justified by the observational fact that wavelength can be photographically recorded and by the analogous facts connecting frequency and wavelength in the case of macroscopic waves. The other extension is the use of a photomultiplier to locate single-quantum absorptions.

Events observed, whether in the focal region or elsewhere, will be given a 'microscopic' coordinate placing (x, y, z, t) in the 'apparatus-frame'. Other events deduced from these by means of a theory will be placed in the 'mathematical frame' which corresponds with the apparatus-frame microscopically. By *microscopic* placing of an event I mean the allocation of spatial measures to it with an accuracy of the order of one micron, 10^{-4} cm. By *submicroscopic* placing I refer to measurements deduced with the help of a further theory and alleged to be of events with spatial separations of the order 10^{-6} cm (100 Å) or less. Quantum theory deals with (alleged) events submicroscopically placed, and these can be loosely described as *submicroscopic phenomena*. In my view, one cannot properly speak of their being 'observed'.

First and second interactions. In quantum physics every observation is presumed to be the immediate result of interaction between submicroscopic structures in the mathematical frame, one of them being anchored in some object in the apparatus-frame (so that its position can be measured). It is particularly to be noted that the alleged interacting structures are not 'observable' in the sense of the term here adopted.

The interaction just spoken of is to be called the *second interaction*. By itself it refers only to the observation in question and the unseen structures from which the observation is supposed to result immediately. The character of the anchored structure is not very important, since it acts only as a detector. The character of the other structure appears when the second interaction is plausibly related to an alleged submicroscopic *first interaction* in another region, which could be previously or subsequently observed microscopically. Hence I shall refer to this other structure as a *connecting* one.

To obtain a 'quantum-mechanical measurement' the connecting structure must be such that one can reasonably suppose a certain property of it (e.g. frequency, velocity) to remain unchanged between the first interaction and the second: also, observation and agreed theory must provide a means of attaching a number to that property in the second interaction. Then this measure is taken to be a measure of the connecting structure established in it as a *result of the first interaction*. Other measures, sometimes excessively small (e.g. electronic charge, Bohr radius), are obtained by the successive application of formulae to *models* conceptually placed in the mathematical frame as if it were the apparatus frame. The effect of the experimental set-up leading to the second interaction thus takes on the aspect of a magnification from the submicroscopic event at the first interaction to a microscopic and thus observable one.

On the other hand, it is more correct to say that the submicroscopic event is 'inferred' from the observation by a complex and abstruse *backward extrapolation*, and that the 'magnification' is thus more strictly an extrapolated *correspondence* between what is observed and unobservable structures postulated in the mathematical frame and obeying mathematical equations. Such correspondence constitutes the quantum-mechanical *explanation* of the observations. It follows, therefore, that for any extension of the scope of explanation we must have, in the first place, new postulated structures, and in the second place, new and satisfactory methods of backward extrapolation to them.

Examples of backward extrapolation. In the case of the photo-electric effect the observation is of the battery voltage and the needle indicating the absence of current passing. It is then necessary to assume classical electrodynamics to deduce that if there are negative particles of a certain mass and charge, released from the metal plate, their maximum kinetic energy must be a certain quantity (whose connection with frequency is then discovered). The claim for the existence of such particles rests on the results of such experiments as Millikan's oil drop experiment and J. J. Thomson's on the deflection of beta rays.† After this, a dubious argument leads to the view that the incoming wave consists of separate wave-packets (photons). [2]

Such an involved dependence on classical concepts and theories is dangerous, and if we accept the conclusion of the Jánossy–Náray [3] and grating experiments, [4] that 'particles' in the free field are unlocalisable separately, then the argument is seen to break down completely.

In contrast, the grating experiment seems to make appeal to nothing but the properties of wave-motion. The conclusion is simply that if there were localised wave-packets approaching the grating the statistical fringe-width obtained on diffraction would be far wider than what it is with an entire wave; and since it is not different, the wave must meet the whole grating however much the intensity is reduced. The further vitally important conclusion is that the 'final decision', so to speak, as to where a quantum is to be absorbed from the entire wave proceeding from the grating is not made till the detecting screen is reached, i.e. the observation is made. The various possibilities are *held in potential*, as we may say, till the actual observation is made. It is specially noteworthy that this view seems to be forced on us independently of the manner in which quantum theory has developed or may develop in the future.

Similar conclusions follow from the Jánossy–Náray experiment (here loosely referred to as an actual two-slit one). It is impossible to trace back a trajectory from the place of absorption to one or other of the two slits where one might suppose the 'particle' passed through. We might consider the possibility that the 'decision' as to how the particle is to move towards the detecting screen is worked out in the region just beyond the two slits, where the waves from them begin to mingle, and that thereafter a fairly clear trajectory is followed, becoming quite precise by the time the detecting screen is reached. But to determine the statistical distribution theoretically (as it is also

† Actually experiment establishes only the discreteness of *absorptions*, not of 'particles in transit'.

actually) we have to carry out an imaginary construction at the screen. It must therefore be admitted that the absorptions occur at the screen in accord with this construction as it is at the screen. The 'particle' must, so to speak, bear in mind this construction at the screen before it can come to any decision as to where it will be absorbed. We merely put this state of affairs in more scientific language when we say that the possibilities are 'held in potential' till the actual observation is made.†

Ontological status of extrapolated events and fields. To see more clearly what this means, let us put the question whether, given a place of quantum-absorption on the screen, we can extrapolate back and allocate the corresponding measure strictly and uniquely to the region just after the two slits. The answer is clearly 'No'. The first interaction creates, in the framework of what is to follow, a *set of possibilities* for an observed measure; and the measure extrapolated on any particular occasion is merely the one that will eventually be obtained on that occasion. It does not exist at the slits except by anticipation and in reference to the whole experimental set-up. Consequently, to get a comprehensive representation of the possibilities just after the two slits, we must extrapolate from many observations made under the same observable conditions.

It may also be recalled that quantum-absorptions in the Jánossy–Náray experiment occur with the normal statistical distribution, even though less than one quantum is in the apparatus at any time (reference (2), p. 316); from which we may infer that the extrapolated field in the mathematical frame, between the two interactions, represents a 'control mechanism' for absorption, rather than a localised transfer of anything specific.

Backward extrapolation thus seems to take us to 'another sphere', of quite different ontological status from that of observations.‡ It is a sphere of *general control-structures* represented in a mathematical frame and not *specific observable* ones in the apparatus-frame. Owing to the way in which the measures are obtained by constructions from the apparatus-frame and with the help of models as if in that frame, the mathematical frame appears as a more detailed duplicate of it. But coincidence of the frames occurs only with observation, and obviously cannot be of more than microscopic accuracy.

† The terms *potential, potentiality* have been used by Heisenberg, Bohm[5] and other authors. Rothstein referred in the colloquium to an object as 'a potential information source'.
‡ Bohm refers to the 'hierarchy of orders' forming the 'living body of natural law'.[5]

Scattering. 'Scattering' is characterised by a spatial spread of absorptions as the result of presumed interaction of a beam or systems with material in a scattering region, or by the correlation of such single absorptions presumably from two beams or systems that have interacted. Since the case of a continuous spread of observed values may be explained as the effect of multitudinous single absorptions, no consideration is given to it here.

More comprehensive information as to the state of affairs just after the first interaction is obtained if we can extrapolate backwards from cloud chamber or other devices which show the timing and direction of momentum $h\nu'$ of each field or system at some stage afterwards. The first experiments designed to give such information were the Bothe–Geiger experiment [6] on the correlated timing of recoil electrons and the slightly later Compton–Simon experiment [7] which I consider here.

It was found that some pairs of electron tracks in a cloud-chamber photograph could be correlated on the supposition that one of the tracks began at the point where a photon (with known momentum) collided with a nearly stationary electron and that the photon then travelled to the point where the other track began, this track being caused by the recoil electron at a second collision. The angles of deflection thus inferred for photon and electron at the first collision were found to be correlated according to the relativistic theory (conservation of 4-momentum) for the system of two simply-located particles.

This result poses a remarkable problem because of the backward extrapolation from two non-simultaneous events. The appearance of the first and immediately following droplets on the first electron-track presumably counts as an 'observation' which actualises the result of the first collision. Thereafter the trajectory of the photon should be decided, as indeed it appears to be, in accord with the calculations. But this means that the track of a photon is definitely decided *before* an absorption from its potentiality field takes place.

Now if we did not happen to observe the track of the first recoil electron, but merely knew the place where the first collision occurred, we would have said, on analogy with two-slit and grating experiments, that the various possibilities for the photon are held in potential till an absorption occurs (Bohr, Kramers and Slater said this in 1924). But in fact the unseen motion of the first recoil electron had determined a specific deflection for the photon trajectory (within certain narrow limits) before any absorption from it took place.

Closer examination, however, reveals fundamental differences between the two cases. In the two-slit and grating experiments, the inferring of a specific trajectory (before or after absorption) is *logically* impossible. In the scattering experiment, the inferring of a photon trajectory is now seen to be conceptually possible, but may be prevented by our lack of knowledge.

Moreover, in the first case the photon is destroyed on absorption; but in the second case the photon remains in potentiality. For a quantum cannot be absorbed from an extended field instantaneously if that field is an actually existing one moving with a finite velocity. Indeed, we have seen that there could be less than a quantum in the apparatus at any time, and still absorption would occur if the incoming (very weak) field were continued long enough. Hence the potentiality field of the photon before the interaction suffers contraction into a narrow cone-like pencil, *while remaining still in potentiality*.

The special feature of scattering experiments is thus the re-organising of the interacting potentiality fields (photon and electron, in the case considered) so that after interaction the total field consists of a set of pairs of correlated pencils, one in each of the constituent fields, but both still in potentiality; and one correlated pair is actualised according to the total experimental conditions.

Quasi-subjective compromise. We have arrived at the conclusion that whereas in the two-slit or grating experiment the inferred potentiality field of light-quanta or electrons collapses in one stage, so to speak, when the absorption occurs, in scattering phenomena it collapses in two stages. At the first (partial) collapse the field remains a potentiality one, more or less restricted, and the occurrence of this partial collapse is postulated because of the accession of new knowledge, from which we are able to extrapolate backwards in more detail than before.

Let us now shift our view to a proposal previously rejected, namely, that the outcome of an experiment is 'gradually worked out', and let us first consider a rough but perhaps serviceable analogy.

Let a lightning flash pass from one cloud level to another, and then on to the earth. There is an electrostatic potential filling all space. The first flash begins presumably where the potential gradient is steepest, other conditions being in order. But before it can have travelled more than a very short way, conditions will have become appreciably altered, partly from external causes, and partly from effects of that part of the flash itself. Each minute step has to be worked out by nature, so to speak, by calculation of the changes in the potentiality field as a whole (now electromagnetic). Thus we can truthfully say that the path is not determined till it actually occurs. Likewise with the later flash; yet the two are correlated in certain respects, e.g. their approximate times.

In quantum theory, when we consider the unobservable region of potentiality to which we try to extrapolate backwards from a few ultimate observations, it seems plain that if there is any such gradual working out, we are, in nearly all respects, *unable in principle* to extrapolate backwards to the details of it. The process is, for the most part, beyond mathematical formulation in principle. The best we can do is to accept a compromise. We introduce a potentiality field which embodies those characteristics to which we can extrapolate, and leave the door open for some refinement (or partial collapse) when some other backward extrapolation becomes possible.

I call this process *quasi-subjective compromise* because the field as we introduce it does embody objective conditions to which we have correctly extrapolated. But we have been obliged to accept a representation of it which is general and possibly incomplete, because of unavoidable limitations to our knowledge. Such representation is not to be considered wrong. Being general, it comprises any narrowed possibility which fuller experience might indicate is the case.

In dealing then with a particular experiment, we may be able to extrapolate backwards to some potentiality state from which the observations may appear to result by sudden collapse or 'actualisation'. Or in some cases, by extending the scope of our observations while the first interaction is reasonably supposed to be the same in type, we may be able to interpose one or two inferred partial collapses, amounting to the same thing in the end.

In this way quantum theory could presumably extend its scope. 'Hidden parameters' might come to light, provided they are, as Bohm requires, such as refer to 'potentialities whose precise development depends just as much on the observing apparatus as on the observed system'.[8] For they must be obtained by some kind of backward extrapolation having the whole set-up in view.

Assuming that some new kind of extrapolation is discovered there could also clearly be justification for modifying the wave-equation (as Bohm and Bub have proposed), in the hope of getting a 'language' which will fit with the new observations—just as the language of field theory, with its creation and destruction of quantum-states, fits better with observations of the Jánossy–Náray or scattering type than does the language of elementary quantum mechanics.

But it is difficult to see how we could ever arrive at a completely specific and accurate formalisation for the continuous process of collapse which it might seem more reasonable to postulate at the back of nature. For such fully formalised explanation would require an infinity of observations (while the experiment is in progress) and an infinity of impossibly complex mathematical problems to be solved as each stage passes continuously, by holistic integration, to the next.

Conditions for backward extrapolation. Now let us consider under what circumstances a backward extrapolation is likely to be possible, and when we must judge it to be impossible. If the observed results at the second interaction fall into some regular order, that is to say, they are the same or similar in some respect when the experiment is repeated and there are elements of continuity, symmetry, or other regularity suggesting a dependence on some mathematical formula, then we may feel justified in looking for an application or extension of some theory which has proved successful in experiments on similar 'materials'. If we are successful, then, corresponding with the observational differences, we choose a suitable functor of the theory with values corresponding to the observed results. The observed result, in any one case, is then said to yield the corresponding measure of that functor ('angular momentum', for example, in the Stern–Gerlach experiment).

At the opposite extreme are observed results that seem completely random. That is to say, they are different when the experiment is repeated, and there seems no regularity either in the total results on any one occasion of observation or between the various results on various occasions of observation (other than what we judge to be irrelevant to the measurement in question). It would seem then that we have no basis on which to attempt any backward extrapolation.

In between come observed results which combine the regular with the random, as in experiments of the two-slit type. The conclusion seems to be that we may hope to explain the regular features by backward extrapolation in terms of some theory, but in respect of the random features backward extrapolation and therefore explanation are impossible. We could carry out a backward extrapolation only by additions to the experimental set-up, by which observations of a new type, exhibiting some order, can be made. The potentiality field is then contracted, but as it is still in potentiality, there will be features (e.g. precise timing) which are still unexplained.

Preparation: 'Filtration'. We must now return to consideration of the experimental conditions, apart from the large-scale and comparatively static parts of the apparatus. The injected materials or fields in the region of the first interaction must all be prepared so that we can be confident that they are of the required character. But here arises a certain familiar difficulty, namely, that the conditions before the first interaction cannot be observed without performing another experiment on them, and so altering them, in some respects at least.

The resolution of the paradox is said to be by the fact that there are measurements which act as *filters*. Thus after passing a ray of light through a Nicol prism we can verify that the transmitted ray is plane polarised. If we pass the transmitted ray through a second Nicol, oriented the same way, we again verify that it is polarised as before. The identity of the conditions after each of the Nicols shows that the second Nicol has no effect on the polarised ray. Hence we can conclude that the ray as it emerges after the first Nicol is polarised, even though no observation is made of it at that stage. Observation of it has not proved necessary.

To set out the argument in this way shows, however, the erroneousness of the usual claim that a 'system' is prepared by performing on it 'the simultaneous measurement of a complete set of compatible variables',[9] as also of the less drastic claim that measurement of one variable A, yielding eigenvalue a', leaves the system in a corresponding eigenstate (this is the Projection Postulate).

For the above illustration shows that one of the features of preparation is that, when the experiment is performed, a measurement is *not* made on either a 'system' or a beam in order to prepare it. No measure is made of the beam as it emerges from the first Nicol; still less (if possible!) is a measure made of a photon in it.

We have to be careful, also, if the beam consists of only one quantum. For consider such a beam or wave-packet approaching the first Nicol. After passing the balsam layer in the prism, the quantum is not to be located either in one beam or the other, unless an observation is made *at that stage*. Hence it is only when observation is made after the second Nicol has been passed that we could be justified in extrapolating backwards and saying that the quantum took the path of the extraordinary ray after passing the first Nicol. Before that, everything is in the region of potentiality, with respective probabilities that absorption might indicate one or the other path.

Rules of inference: The Projection Postulate. From the above analyses it is clear, I hope, that in place of the usual Projection Postulate (reduction of the state-vector) one needs very carefully framed Rules of Inference or definitions. I suggest the following as covering the essential phenomenological basis, free of inadmissible or needless presuppositions.

RULE 1. Let an experimental set-up consist of materials and fields which have been prepared by a prescribed procedure and remain unchanged during a certain period (called the period of the experiment) or at least are the same on each occasion of observation during that period. Let this set-up provide 1st and 2nd interactions, each in a given microscopic region, and an inferred connecting structure. Let various 2nd interactions result in observations by virtue of which measures $a_1, a_2 \ldots$ of a functor A are each associated with an inferred connecting structure at the 2nd interaction. Corresponding state-vectors of the functor A are then to be *conjointly allocated in potential* to the connecting structure just after the first interaction, and the structure is to be denominated a *ray*.†

RULE 2. If, when a ray has been produced as in Rule 1, we omit those parts of the set-up which provided the 2nd interaction, we may nevertheless postulate the unchanged existence of such connecting structure and denominate it a *prepared ray*. State-vectors corresponding to $a_1, a_2 \ldots$ are still to be conjointly allocated in potential to it.

RULE 3. If a prepared ray enters an experimental set-up, and, as a result of observations at various 2nd interactions differently placed microscopically or macroscopically, various connecting rays from a 1st interaction are inferred, then the set of such rays, regarded as a continuing whole, is to be denominated a *beam*. State-vectors corresponding to the various measures $a_1, a_2 \ldots$ are to be conjointly allocated in potential to it.

After this one can introduce rules for inferring a *locale* or interacting *system*, whose continuing character is to be inferred by means of

† For example, the emission of a ray of light from an incandescent grain of common salt. The state-vectors are sinusoidal in the various frequencies, and in the field-theoretic formulation can be of 0, 1, 2 ... quanta.

various connecting structures.† It must be represented by a ψ field in some localised way so as to accord with a theory of quantum-interchange. Such *locale* or *system* is a potentiality structure extrapolated from observations, like rays and beams, but located in a materially identifiable microscopic region (I see no way of defining a 'system' in a ray or beam, or in fact any *submicroscopic* system).

DEFINITION 1. A prepared ray or beam is to be called *interferable* if it is possible in some way to divide it in two at a 1st interaction (this division being inferred by backward extrapolation) and to superpose the two derived rays or beams in a region of 2nd interaction, so that a statistical distribution in bands is exhibited, indicating destruction of the potentiality in certain places.

DEFINITION 2. A prepared ray or beam is to be called a *mixture* in respect of functor A (or a mixture in A) if the statistical distribution of the various measures a_1 yielded can be treated as the superposition of two or more statistical distributions, or otherwise if it is not interferable.‡

DEFINITION 3. An interferable ray or beam which is not a mixture in any functor is to be called *simply interferable, coherent,* or *pure.*§

The next step might be to elaborate the definition of the state-vector for a pure ray or beam as the linear superposition of the separate state-vectors allocated to it, with coefficients c_1 such that $|c_1|^2$ are the classical probabilities for the respective measures. For reasons of space, further steps in the development are omitted here.

As a simple illustration, consider the well-known Einstein–Podolski–Rosen paradox. There it is assumed, not only that a two-fold system possesses certain specific measures actually, and that the system can be separated while the parts retain their particular measures, but also that various second interactions may be arranged so that after *these* interactions the combined system assumes eigen-

† A state-vector suitably imputed to a microscopic grain of NaCl would be an example of a *locale* or *system* in this terminology. It cannot be imputed to any particular molecule, but applies to the structure in general.

‡ Example: A ray with frequencies ν_1, ν_2 from different monoenergic sources superposed, or a ray that has travelled so far that the phases have become confused (as deduced from its non-interferability). Every beam represents imperfect knowledge, in so far as we do not know where or when an absorption from it is going to take place. A mixture is no better and no worse than a pure beam in that respect.

§ Note that a ray with frequencies ν_1, ν_2 from different sources is normally interferable, but is also a mixture.

values of the corresponding operators, such eigenvalues being then identified with the measures supposedly possessed after the first interaction. Numerous explanations of the E.P.R. paradox have been offered, of course, during the last 32 years. But nearly all have accepted the three authors' application of the Projection Postulate; and this seems to show that the fact that *two* interactions are involved in any measurement process was not sufficiently realised, nor the fact that, in general, we cannot impute a specific character to the state of affairs immediately after the first interaction; we can only declare the measure after and in accordance with the particular second interaction set up.

This lack of clarity seems to show also in the *measurement algebra* recently developed by several authors, in which $M(a_1)$ corresponds to 'the process that selects systems that have the value a_1 of A'.[10] Although a single value a_1 can be rightly attributed to an appropriately prepared ray, it is to be noted that in the event of two or more values a_1, a_2 being possible, the ray or beam cannot be said to possess definitely either the one value or the other. Hence to say that '$M(a_1) + M(a_2)$ corresponds to a filter that accepts systems that have *either* the value a_1 *or* the value a_2' is to describe the situation inaccurately. It would be more correct to say that '$M(a_1) + M(a_2)$ corresponds to a ray, beam or system to which the values a_1 and a_2 must *both* be allocated in potential'.

General conclusions. To sum up, I should like to suggest the following as being particularly important for any discussion of the 'beyond' in quantum theory:

(1) Any formalism which makes use of the Projection Postulate is phenomenologically misleading, and conclusions drawn from it cannot be depended on.

(2) There is no phenomenological warrant for breaking up rays and beams into separate 'particles' or 'systems', and the proposal to do so is inadmissible at least when the rays or beams are interferable.

(3) There is no reason why a better 'language' of state-vectors and for changes at the 1st interaction, making use perhaps of new kinds of backward extrapolation, should not be discovered, provided the distinction between classical statistics for mixtures ($|\psi|^2$ additive) and quantum statistics for pure rays, beams and systems (ψ additive) is maintained.

REFERENCES

(1) Seitz, F. Solid state physics, *Enc. Brit.* (1963).
(2) Whiteman, J. H. M. *Philosophy of Space and Time* (London, 1967), 314.
(3) See reference (2), 316.
(4) Messiah, A. *Quantum Mechanics*, vol. 1 (English translation, 1964), 19.
(5) Bohm, D. On creativity, *Leonardo*, vol. 1 (Pergamon Press, 1968), 141–3.
(6) *Z. Phys.* (1925) **32**.
(7) *Phys. Rev.* (1925) **26**, 289.
(8) *Phys. Rev.* (1952) **85**, 187.
(9) See reference (4), 204.
(10) Gottfried, K. *Quantum Mechanics*, vol. 1 (1966), 193.

MEASUREMENT THEORY AND
COMPLEX SYSTEMS

M. A. GARSTENS

Introduction. The problem of measurement in atomic physics belongs to the discipline of statistical mechanics. Like the latter, measurement theory involves the attempt to weave together a consistent picture of the microscopic and the macroscopic aspects of the world. Although all observations are macroscopic in origin there is in quantum theory an undeniably accurate picture of the sub-structure underlying everyday macroscopic experience. However, there never has been established a clear link between the accepted quantum mechanical sub-structure and the source of information about it, which is macroscopic in origin. In the past the macroscopic has been the subject matter of classical physics. There appears to be a gap between the theory of the macroscopic and the theory of the microscopic.

The attempt to bridge this gap may simultaneously resolve some of the difficulties present in fundamental particle research. In addition there are indications that the resolution of the measurement problem will have far reaching impact in clarifying other areas of research. In particular it is probable that the removal of the barriers to the attainment of a more satisfying theoretical outlook as to the nature of complex processes in fields like biology, also awaits the solution of the measurement problem. This is due to the fact that in biology disturbances due to measurement of basic parameters play as important a role as in the measurement of atomic variables.

1. The Problem of measurement. If measurements of a dynamical variable A are made in a large number of identical systems each with wave function before measurement:

$$\Psi = \sum_i c_i \psi_i$$

then one observes in each successive system some eigenvalue a_i associated with the wave function ψ_i with probability c_i. The selection by a single system of its state ψ_i (i.e. $\Psi \to \psi_i$) is known as the collapse or reduction of the wave packet. In practice one observes the distribution of the different c_i and then matches them with the assumed distribution indicated by the undisturbed wave function Ψ. The problem is how Ψ goes over to one of its single components. Is the process stochastic or do the accepted quantum mechanical laws play a role during the reduction?

The answers given to this question, of which there are many, are sensitively dependent on one's conception of the nature of quantum theory or of physical theory in general. Acceptance of the stochastic interpretation, as by Bohr [1], implies the impossibility of determining physical processes beyond a certain limit (defined by the indeterminacy principle). Thus he assumes that macroscopic observation, in the sense of classical physical theory, is primary. Quantum theory is for him an algorithm, a set of rules telling how to calculate, using observed data as a starting point. The status of the reality of the algorithm or set of rules, a subject we will comment on later, is left a little unclear. There is an old controversy as to the reality of universals which has much bearing on this point. It is difficult to accept the view that a successful theory, as quantum theory is, should be deemed to have less or a different kind of reality, than the macroscopic sources from which it stems. Using his interpretation, Bohr avoids the problem of wave collapse, since to him, it represents the observed fact of irreducible randomness in nature. It is however of utmost importance to realize, contrary to the position of von Neumann, [2] that a denial of the validity of the present formalism (or algorithm) of quantum theory, as may be involved in hidden variable or other types of formulations, does not necessarily imply a denial of the existence of randomness in nature or of some form of indeterminacy. The limits and meaning of randomness can be radically altered by alternative formulations. Of course experimental observation should dictate the need for such new formulations.

In von Neumann's [2] approach to the problem of measurement, the stochastic reduction of the wave packet is analysed in causal terms, using the quantum equations of motion. His solution is unsatisfactory since it demands that each measurement be observed by a subsequent one, leading to an infinite series which can only be resolved in a final conscious act of observation by a person. This subjective interpretation is not acceptable since we know that observations are not subjective and do not require this endless series. It is of interest to note that in spite of much subsequent criticism of von Neumann's approach, Wigner [3] contends that no adequate substitute for von Neumann's analysis of the problem has thus far been presented.

Bohm and Bub[4] have attempted the use of hidden variables to attain an explanation of wave collapse by more deterministic methods than current quantum mechanics can supply. The experimental check of this approach is still pending. The paucity of experimental verification of substitutes for quantum theory has in fact been their primary weakness. One wonders in fact whether any complete physical theories have ever been posed in advance of knowledge of the experimental situations which require them. We will suggest, in this connection, that the direction of the 'beyond' toward which this conference on quantum theory would like to travel, could be determined by an existent rich body of empirical data requiring explanation. These data are in the field of biology, the domain of the complex.

2. Reality in physics. It is one of the strange phenomena in modern physics to observe an increased concern with defining the reality which theory attempts to describe. Such discussions (to many physicists undesirable) are a sure sign of the unsettled nature of current theory. Thus in a paper by Jauch, Wigner and Yanase [3] it is stated that:

The concept of a 'physical reality' as far as inanimate objects are concerned, may itself lack of 'physical reality' just as the concept of absolute rest does. What we can do is to foresee, to some degree, what we are going to experience and all other questions concerning 'reality' may constitute only an unnecessary superstructure. In the terminology of K. R. Popper, the reality may not be 'falsifiable'. This does not mean that we have to abandon the concept of 'reality' altogether. It only means that this concept does not seem to be necessary for the formulation of the conclusions to which physical theory leads us.

In the well known paper by Einstein, Podolsky and Rosen on 'Can quantum-mechanical description of physical reality be considered complete?', they say: [5]

A comprehensive definition of reality is...unnecessary for our purpose. We shall be satisfied with the following criterion, which we regard as reasonable. If, without in any way disturbing a system, we can predict with certainty (i.e. with probability equal to unity) the value of a physical quantity, then there exists an element of physical reality corresponding to this physical quantity. It seems to us that this criterion, while far from exhausting all possible ways of recognizing a physical reality, at least provides us with one such way, whenever the conditions set down in it occur. Regarded not as a necessary, but merely as a sufficient, condition of reality, this criterion is in agreement with classical as well as quantum-mechanical ideas of reality.

Again in the various discussions of hidden variables [4] the question of the nature of physical reality is of primary importance. The claim that quantum theory is lacking in completeness of description means that important portions of reality are missing. The 'algorithm' of Bohr is something of 'lesser' reality in his interpretation. On the other hand it is frequently felt that the atomic 'underpinning' of things has greater reality than things directly observed. This undoubtedly has been a strong reason for the attempt, in the reductionist philosophy, to explain everything in terms of atoms.

It will be noted in all of the above discussion that the notion of reality, no matter how defined, has an ineradicable element of the subjective in it, involving a choice, a value judgment, or a decision as to what experience should be reduced to what other experience, a conclusion as to which domain is more real. The decision to attempt the reduction of observed phenomena to their atomic constituents, a subjective act, is made because of the richness of discoveries and predictions to which it leads. But logically, even if with great difficulty, the atomic can be deduced from the macroscopic. If there were a loss of interest in discoveries and predictions, for whatever reasons, then a shift could occur where the macroscopic would be considered more real and reduction could go the other way. In fact the possibility of development of a science of the complex (i.e. of the theory of biology) would be enhanced if such a shift were to take place. Logically this means the development of the mathematical structures which more directly describe observed complex mechanisms and which are considered more central than the theories of the underlying microscopic structures already developed. This would not eliminate the need, in so far as microscopic effects are present, for adequate mathematical theories to take them into account. The programme of reducing these complex areas to current atomic theory, which is the 'style' of our times, cannot be carried out because of the complexity of the wider areas of experience which must be understood. There is thus a question of value involved here, social or individual, in which a decision must be made as to which is the more important reality *now*; which studies most deserve our present attention.

3. Statistical mechanics. To bolster the above remarks, it must be shown that complex macroscopic structures are as basic as their atomic substructure. The whole ergodic programme in statistical mechanics is based on the denial of this contention. The difficulties in the ergodic programme have not been resolved. [6] They have led to the observation that in order to successfully deduce macroscopic properties from current statistical atomic theory, supplemental special assumptions or hypotheses are required.[7] It is probable that the special auxiliary assumptions needed in each case depend on the specific macroscopic properties to be explained. Important to note is the main contention being made here, that these auxiliary hypotheses are as basic (and as real) in these contexts as the rest of atomic theory. They act as a type of boundary condition, and point to the need of pluralism in hypotheses, reflecting the pluralism in nature revealed in the more complex areas of our experience. A thorough-going reductionist programme would thus seem to be the wrong direction for physics at the present time. This broader pluralistic point of view encourages direct logical or mathematical analyses of biological phenomena as being realistic and scientific. The connections with the atomic domain are to be supplied by the discovery of auxiliary assumptions playing the same role as they do in statistical mechanics. As in statistical mechanics, the greatest advances can be expected when these assumptions are realistic and productive. The great art is in finding such assumptions.

The upshot of the above remarks is perhaps similar to the message of those who advocate the reality of hidden variables. The real motivation in such a programme, as presented by Bohm,[8] is to get around the rigidly reductionist outlook which the presently accepted interpretation of quantum theory presents. The axioms for quantum mechanics, presented by von Neumann, seem to have placed an almost iron-clad boundary about the field. The point of view presented here and that involving the introduction of hidden variables emphasize three essentials: the impossibility of any scientific theory ever completely defining the essence of nature, the pluralism manifested by the complexity of nature, and the need to broaden the base of the scientific theory in order to see the multiplicity of nature in an increasingly unified way.

REFERENCES

(1) Bohr, N. In *Albert Einstein: Philosopher-Scientist* (editor P. A. Schilpp; 1949).
(2) von Neumann, J. *The Mathematical Foundations of Quantum Mechanics* (Princeton, New Jersey: Princeton University Press, 1955).
    Also:
    London, L. and Bauer, E. *La Theorie de l'Observation in Mecanique Quantique* (Paris: Hermann et Cie, 1939).

(3) Jauch, J., Wigner, E. P. and Yanase, M. M. *Nuovo Cimento* (1967) **48**, 144.

(4) Bohm, D. and Bub, J. *Rev. mod. Phys.* (1966) **38**, 470.

(5) Einstein, A., Podolsky, B. and Rosen, N. *Phys. Rev.* (1935) **47**, 777.

(6) Farquar, I. E. *Ergodic Theory in Statistical Mechanics* (Interscience Publishers, 1964).

(7) Ludwig, G. *Axiomatic Quantum Statistics of Macroscopic Systems in Ergodic Theories* (Academic Press, 1961).
     Van Kampen, N. G. *Fundamental Problems in Statistical Mechanics* (editor E. G. D. Cohen; 1962).

(8) Bohm, D. This conference.

IV

NEW DIRECTIONS WITHIN QUANTUM THEORY: WHAT DOES THE QUANTUM THEORETICAL FORMALISM REALLY TELL US?

The two parts of this book to follow contain papers that are written from a standpoint critical of the adequacy of the current foundations of quantum theory. The authors of the papers in the first of these parts—the present—all start from the existing mathematical structure of quantum theory and discuss its range of applicability in different ways. This question of the range of applicability of the formalism is really upon us as soon as we look critically at Bohr's assumption of the inevitability and immutability of the whole language of classical physics as the interpretative vehicle for the formalism. A position favoured quite widely in the colloquium was to accept the thesis that classical ways of thinking are indeed needed at every point to interpret quantum-theoretical formalism, and yet hold that a full understanding of the quantum theory should or could provide us with a fuller and essentially more correct set of concepts for describing the physical world. The classical language—on this view—is not immutable but is subject to alteration, refinement and development.

Expressing one extreme case of this view in his paper, Kilmister suggests that quantum theory may amount to little more than the imposition of a set of combinatorial conditions upon the energy-values in an essentially classical picture, and that its claim to be a new system of mechanics in its own right is overstated if not actually misleading. His central argument in support of this view is that the test of quantum theory having really generated concepts properly its own should be the power of the theory to guide our thinking in circumstances in which the classical ideas are known not to apply, and he finds that the theory does not pass this test.

Bohm and Aharonov—in different ways—find the potentiality of the quantum theoretical innovations to be greater. During the discussions Aharonov insistently posed the question which heads this part of the book. The quantum mechanical formalism—he contended —does not have exactly the same scope as the classical language of physics. On the contrary, it leads to quite characteristically quantum mechanical ideas. These latter—contrary to what is usually thought— have not been assimilated by physicists because the full possibilities allowed by the formalism have not been exploited. Aharonov's actual paper with Petersen is less general and discusses whether the 'scope' of the quantum formalism extends to the relativity principle.

Bohm's position was different. He did not think it possible or desirable to make a sharp division between mathematical formalism and the ideas which spring from the formalism or which the formalism was invented to express. The dependence of quantum theory upon classical ideas and classical language is something which Bohm is ready and even anxious to see modified by the development of new forms of expression. He discusses a new approach to the spinor calculus which might be a first step in such a development. Bohm's paper is followed by a section of the discussion to which it led. This piece of discussion concentrated most of the important points that emerged in the discussion of this section in a short space, and it seemed a good thing to preserve it as it stood.

The attempts of Bohm and Aharonov to embed the quantum formalism in a richer and characteristically quantum mechanical set of ideas and language was associated in its origins with the careful scrutiny of quantum mechanics for the possible existence of hidden variables. What inference should be drawn from the von Neumann proof of the non-extensibility of quantum mechanics (in the sense of the inadmissibility of 'hidden variables') has always been in some doubt, particularly in view of difficulties (of a sort that became very evident in this colloquium) in completing quantum mechanics by the addition of a quantum mechanical theory of the measurement process. Is quantum mechanics itself a complete theory? This question was a major issue in the discussions, for hidden variable theories tended to have an essentially retrogressive feel about them since they often provided no place for the new insights of quantum theory. In the discussions in the colloquium it was allowed to go without saying that any acceptable hidden variable theory would have to avoid this defect. In fact, the participants in general clearly felt that a definitive picture of the outcome of the 'hidden variables' controversy was emerging.

The discussions and papers at the colloquium scarcely touched on technical elementary particle theory, nor even on those extensions of quantum theory which, like quantum field theory, have been put forward to provide a theoretical basis for the understanding of the elementary particle. This omission was dictated partly by the spheres of interest of the participants, but it also reflects an uncurious attitude towards the conceptual difficulties of quantum theory at the foundational level on the part of those engaged in particle physics. The omission, indeed, is not altogether satisfactory. Many people who work in the particle physics field find themselves forced to make judgments which depend upon the view they take of the quantum theoretical foundations, and obfuscation can often be produced if these foundational judgments are settled at a more superficial or technical level than that to which they really belong.

Occasions when judgments of this sort become important arise, for example, through efforts to extend relativistic invariance to the whole quantum theoretical formalism, particularly in cases when that formalism has to be applied to high energy particles, and the colloquium discussed this issue extensively. It is well known at the technical level that no relativistic particle theory is completely adequate (though how serious the troubles in this direction are, is a matter for dispute). It is also known that the really foundational questions of quantum theory are discussed without reference to, and against a quite different operational background from the relativistic concepts. Some participants wanted to hold open the possibility that the technical inadequacy was connected with the foundational one.

Aharonov and Petersen, in their paper, survey the field that quantum theory is commonly claimed to cover, in terms of a concept—'definability'—which they introduce. They are concerned about the varying operational backing that exists for concepts that can be defined by the theory. They conclude that in this respect non-relativistic quantum theory is satisfactory but that relativistic theory is not.

Chew, in his paper, continues this theme more closely in the context of the physics of elementary particles. The S-matrix philosophy attaches primary significance to a certain formal structure—namely the S-matrix itself—transformations of which manifest themselves as particles. The particles accordingly have a secondary role in that theory and one can no longer take them as elementary in the logical sense. With analytic continuation, the matrix-transform method has ceased to be a phenomenologically economical device for operating with the energy levels of atomic structures which are assumed to be there in the background: it has undertaken to explain the origins of the structure.

In these respects the S-matrix approach constitutes a far-reaching change at a level which was within the terms of reference of this colloquium. To discuss it at that level was felt to be taking on too big an issue and—as with papers which make reference to the phenomenon of life in the last part of this book—all the colloquium could do was to register interest in a field of enquiry of potentially great interest to it. At the more technical level, therefore, Chew discusses the fundamental point at which he considers analytic S-matrix theory to depart from conventional quantum theory—namely in the assumption of the completeness of Hilbert space. Hilbert space completeness is relevant to finitism and to the definition of order of points in space or time which play a large part in later papers (and in the discussion we print following Bohm's paper). However Chew associates the departure with the difficulties of formulating quantum theory relativistically.

ON THE ROLE OF HIDDEN VARIABLES
IN THE FUNDAMENTAL STRUCTURE
OF PHYSICS

D. BOHM

1. Introduction. Ever since the quantum theory took its modern form, determined to a large extent by Bohr's notions on the wholeness of the experimental phenomena and on the need for mutually incompatible but complementary descriptions of these phenomena, the possible role of a formulation of the laws of physics in terms of as yet 'hidden' variables has been obscure and, in some sense, confused as well.

On the one hand, Bohr's views on the meaning of the term 'quantum' appear to imply at the very outset that the term 'hidden variables' has no conceivable place in the quantum theory at all. That is to say, as Einstein ruled out the terms 'absolute space' and 'the ether' as irrelevant, so Bohr's intention was to rule out as irrelevant the whole structure of language in which the words 'hidden variables' would be able to have the kind of dynamical significance that physicists generally attributed to them. Thereafter, to try to 'disprove' hidden dynamical variables, or 'show their impossibility' would be a meaningless effort, just as it would be pointless as well as confused in Einstein's language to try to 'disprove' the existence of absolute space and time.

On the other hand, a rather different line of approach was adopted by a number of other physicists, which culminated in the work of von Neumann. In this work, there appeared a well-known theorem which was regarded as a *proof* that hidden dynamical variables are impossible. At first sight, one might be inclined to suppose that such a proof must be in complete agreement with Bohr's basic aims and, in fact, a kind of additional 'support', helping to establish the validity of Bohr's point of view. But if one thinks more carefully, òne sees that this is not the case at all. For if von Neumann's point of view was such as to allow a theorem aimed at disproving the existence of hidden dynamical variables to be formulated, this implies that his language gave meaning to the term, but simply demonstrated logically that this meaning was incompatible with certain axioms or experimental facts that have been accepted as true. But Bohr *begins* with a language that rules out these terms as meaningless. Thus, the two views are not only incompatible; even more, it is a source of confusion in Bohr's language merely to try to talk about certain notions regarded as basic in von Neumann's theory. One sees thus, on reflection, that Bohr and von Neumann were not even related enough in their points of view to meet and to argue the question of which was better.

It does not seem to have been widely realised that there was such an unbridgeable chasm separating the fundamental ideas of the leading physicists who had developed the quantum theory and who were continuing to work on it. Rather, it was generally believed that all were in basic agreement, except for what were thought to be relatively superficially different forms of the language in which the theory was expressed. Since this difference in language was in fact crucial rather than superficial, it was inevitable that there would arise a deep kind of confusion concerning the general informal and 'philosophical' questions, having to do with the meaning and interpretation of the theory as a whole.

This confusion could escape detection, at least in a large measure, because all talked in terms of the same formal algorithm, which was indeed very widely taken to be the 'essence' of the theory. Since such widely different 'philosophical' and 'linguistic' notions seemed to lead to no difference in this 'essence', people could easily come to the conclusion that the general informal discussion of the basis of the quantum theory was itself not very significant. All that seemed really to count was the formalism, with its ability to yield a very wide range of predictions that were in agreement with experiment.

But of course such conclusions only served to add to the problem, because one's overall thinking about the theory is an indivisible whole, in which confusion in one's informal and philosophical premisses will in fact lead to a corresponding confusion in what one does with the rest of the theory (especially when one tries to understand new domains of phenomena where new lines of thought are called for). And because the relevance of such considerations was not widely understood, quite contradictory lines of reasoning continued for many years to enter into most discussions of the foundations of the theory. When these contradictions led to difficulties, it was possible apparently to remove them, either by saying that they had no *formal* or *practical* significance (thus using the tacit notion that what is informal and 'philosophical' has no basic relevance) or else by saying that they were *problems*, that could be solved later (e.g. when the task of reconciling Bohr's and von Neumann's views is called a problem, one can cease to be aware that it is still a contradiction and, in reality, impossible).

In the background of this kind of general lack of clarity and thinking about the meaning and interpretation of the theory, it was inevitable that the reason for entertaining the notion of hidden variables would also not be clear. Indeed, the consideration of hidden variables was very generally regarded as nothing more than a 'reactionary' step, aimed at reintroducing the mechanical determinism of classical dynamics into the basis of physics. In addition, it was felt to be an effort to overturn Bohr's fundamentally new approach that ruled out as irrelevant the old classical ideal of a purely dynamical description of phenomena. In this connection, Einstein's criticisms of the quantum theory as an 'incomplete description of reality' may well have helped to create a climate of thought in which one could easily assume that hidden variable theories must necessarily be intended to provide a more nearly 'complete' description, that would in some sense be an extension of the classical dynamical notions of a detailed analysis of the world into mechanical parts, following well defined laws of notion.

In this paper, it will be my purpose to discuss these questions, and to show that the consideration of hidden variables need not have 'reactionary' aims of the kind described above. Rather, as will be brought out, its deep intention is actually the continuation and extension of what is most basic and novel in Bohr's insights. It is only because the term 'hidden variable' was tacitly restricted to mean 'essentially classically dynamical' that one could plausibly suppose hidden variables to be incompatible with what is new in quantum theory. Indeed, it will be my aim to show in terms of a specific illustrative example, involving the use of hidden variables of a new non-dynamical kind, that this notion can actually help in a significant manner to lead our thinking away from classical concepts and modes of description toward radically new notions of the form of physical law, and even of what is the nature of the activity that constitutes physics itself.

2. On von Neumann's theorem concerning hidden variables.

Von Neumann begins the formal part of his treatment of the subject[1] by enunciating a number of axioms which he regards as basic to the quantum theory. We need not go into these axioms in detail here because, for the purposes of the present discussion, it will suffice to call attention to a few key points:

1. He introduces the term 'observable' in a role which suggests that he takes this to be primitive, i.e. not definable or analysable in detail.

2. He assumes that for each observable, A, there is a linear operator (A).

3. He introduces the notion of the average value of an observable, $(\bar{A})$, to be determined in terms of the operator (A) in a suitable way which he specifies in detail.

4. He assumes a certain further linearity property for this average,

$$\overline{(aA + bB)} = a\bar{A} + b\bar{B}$$

where a and b are arbitrary constants.

From von Neumann's mode of presentation of these assumptions and notions (along with others not mentioned here), we can see that he feels that they lead to a theory which is in essence in agreement with what physicists have generally been using *informally* in their actual work in the subject that is called 'quantum theory'. However, because the axioms have now been *formalised* in a relatively precise way, it is possible to deduce certain theorems, and one of these is that there can be no 'dispersion-free' ensembles (i.e. ensembles that are free of statistical fluctuations for all observables).

To see the relevance of this theorem for hidden variables, it is useful to consider the subject of statistical mechanics, which in a certain sense can be said to provide an explanation of thermo-dynamics in terms of atomic variables that are 'hidden', at least in the context of thermodynamics, as well as that of macroscopic dynamics more generally. One introduces classical 'observables' such as $A(p, q)$ and $B(p, q)$, which are functions of the positions, q, and the momenta, p, of all the particles constituting a thermodynamic system. Then, in terms of the probability density $\rho(p_1, q_1)$, in phase space (with element of volume $d\Omega$) one has for the average of an 'observable'

$$\bar{A} = \int \rho(p_1, q_1) A(p_1, q_1) d\Omega.$$

Evidently from the above definition, one can obtain von Neumann's linearity property

$$\overline{(aA + bB)} = a\bar{A} + b\bar{B}.$$

One then says that there are certain 'normal' distributions, given by

$$\rho = \alpha e^{-E/kT}$$

which describe the situation in thermodynamic equilibrium. In addition, there may be other 'non-normal' distributions, of which a special case is the 'delta-function' distribution, that is 'dispersion-free'.

It is clear from the above that if the 'hidden variable explanations' of quantum theory are to be anything like atomic theoretical explanations of thermodynamics, then they must, in principle, be capable of dispersion-free ensembles. So in a general sense one sees already that von Neumann's axiomatic formalisation of the quantum theory is not compatible with anything so obvious and direct as to make a quantum-mechanical observable come out as an average of a corresponding function of 'hidden' dynamical parameters.

However, von Neumann's theorem has aspects that are more subtle than this. Indeed, he shows that his axioms imply the existence of a *statistical matrix*, (ρ), which plays a role somewhat analogous to the classical probability density, ρ, in the sense that the average of an observable is given by the formula

$$\bar{A} = \overline{(\rho A)}.$$

The linearity postulate is now replaced by the simple expression

$$\overline{aA + bB} = \overline{(\rho(aA + bB))} = a\overline{(\rho A)} + b\overline{(\rho B)}.$$

Or in terms of matrix elements, it becomes

$$\overline{aA + bB} = a \sum_{mn} \rho_{mn}(A)_{mn} + b \sum_{mn} \rho_{mn}(B)_{mn}.$$

So we see that what is common to classical statistical mechanics and quantum theoretical averages is that they are composed of products of *statistical functions* (indicated by the symbol ρ, or (ρ)), and *observational functions* (indicated by the symbol A, or (A)).

How can these relationships be considered in terms of hidden variables? To do this, let us introduce a general symbol λ, representing as yet unknown sets of hidden variables. In addition, there will be further 'non-hidden' variables, representing what is now observable in the quantum theory. Since a 'complete measurement' is now supposed to determine the 'quantum state' which is in turn determined (except for an arbitrary phase factor) by the wave function, it seems natural to take the 'non-hidden' variables to be the coefficients c_r of the wave function in an orthonormal expansion,

$$\Psi = \sum_{r} c_r \Psi_r.$$

In the spirit of classical mechanics and of von Neumann's formulation of the quantum theory, let us define a corresponding statistical function $f(\lambda, c_r)$ representing the probability distribution as a function of the hidden variables λ and of the 'non-hidden' variables c_r. We write $\rho_{mn} = \rho_{mn}(\lambda, c_r)$ and obtain

$$\bar{A} = \sum_{mn} A_{mn} \bar{\rho}_{mn}$$

where $\bar{\rho}_{mn}$ is the result of averaging $\rho_{mn}(\lambda, c_r)$ over λ and over all the c_r. From this one evidently realises the linearity postulate

$$\overline{(aA + bB)} = \sum_{mn} (aA_{mn} + bB_{mn}) \bar{\rho}_{mn}.$$

We can then suppose that there is some 'normal' distribution $f_N(\lambda, c_r)$ (which leads to the usual quantum mechanical averages) as well as a set of more general 'non-normal' distributions, including 'delta functions' as limiting cases. And if $f(\lambda, c_r)$ is a delta function, then all results are evidently determined without dispersion. Therefore, von Neumann's axioms leading to the impossibility of dispersion-free ensembles, are incompatible with this kind of explanation of quantum theory in terms of hidden variables.

A little reflection shows, however, that the above class of theories of hidden variables, ruled out by von Neumann's theorem, is actually a very limited and restricted one. What is essential to this class is that all averages be *linear* functions of the matrix elements, A_{mn}.

One can easily consider more general relationships involving non-linear functions of the matrix elements A_{mn} such that

$$\bar{A} = \overline{\rho(\lambda, c_r) U(\lambda, c_r, A_{mn})}.$$

One can then suppose that there is some 'normal' distribution, in which the above reduces to the relationships applying in the usual quantum theory, while for 'non-normal' distributions, the results will be basically *different*, (i.e. what is now *called* 'quantum theory' will no longer be relevant).

Evidently such non-linear structures of theory go outside von Neumann's linearity postulate, and are therefore not ruled out by his theorem. Indeed, in the next section, we shall discuss a specific example of such a theory, based on non-linear rather than linear relationships.

Thus far, the significance of von Neumann's theorem seems to be clear enough. But then, there appears to exist a widespread belief, perhaps more tacit than explicit, that somehow von Neumann's theorem had ruled out *all possible kinds* of hidden variable theories. How could such an impression have arisen?

If one looks into von Neumann's discussion of the significance of his theorem, one sees in it a number of statements which are ambiguous enough to allow of this latter interpretation. For example, he says that in order for hidden variable theories to be applicable, the quantum theory would have to be *objectively false*. What does this mean? It can be taken in two ways:

(1) Hidden variables imply that the quantum theory has to be *objectively falsifiable*.

(2) Hidden variables imply that the quantum theory would already have to be *objectively falsified*.

The first of these interpretations is evidently a trivially true one. Thus, the existence of 'hidden' atomic variables implies the *potential falsifiability* of thermodynamics (e.g. the existence of Brownian motion and other fluctuation phenomena). On the other hand, since there are no experiments that already falsify quantum mechanics, the second interpretation implies the actual impossibility of hidden variable theories. But, of course, this interpretation does not necessarily follow from von Neumann's theorem, since it could always be supposed that in experiments done thus far, the distribution had not deviated significantly from the 'normal' one, that leads to the usual quantum theoretical results.

One may be inclined to wonder what could have led von Neumann to state his theorem in this rather unclear and even misleading way. A significant clue is perhaps provided by considering the role of the commonly accepted notion that '*basic*' physical theories should in principle provide *complete* descriptions of the world. Thus few people regard thermodynamics as a *basic* theory, and therefore there was comparatively little resistance to the notion that statistical mechanics could perhaps imply the possibility of falsifying it. But the notion of potential falsifiability of a theory regarded as basic in the above sense can be both disturbing and confusing. Indeed, if the terms of the theory are assumed to provide a *complete* description of everything, what can its potential falsifiability possibly mean? Logically speaking that which led to the falsification would have to be *undescribable*. In other words, the language of physics would not be capable of expressing the potential falsifiability of the quantum theory. Now such a position is unacceptable and therefore we must rule out this possible interpretation of von Neumann.

When, however, we cease to regard the language of *any* theory as a 'complete' description of reality, the potential falsifiability of quantum theory is no longer an 'undescribable' thing, and is therefore also no longer tacitly ruled out as irrelevant. Thus we see how it can make sense to go on to the next stage and formulate a specific example of a hidden variable theory.

3. An illustrative example of a non-dynamical hidden variables theory. Recently a theory of hidden variables has been proposed by Bub and the writer,[2] which implies a non-dynamical approach to the conceptual structure of physics, that is in many ways very different from that of classical mechanics.

It was not expected that our 'hidden variable' theory would give detailed numerical predictions that could be compared with experiments. Nevertheless it can reasonably be hoped that it would serve as a kind of 'bridge' toward further theories, that may ultimately yield a relatively detailed kind of content which could perhaps be relevant even for making fairly precise experimentally testable predictions of a numerical character.

3.1. Preliminary discussion. To bring out the full context in which our proposals are to be understood, it is useful to go briefly into what has been called the 'measurement problem' in quantum theory. This problem arises to a large extent because of a widespread tendency to regard Schrödinger's equation as a kind of extension of dynamics, which is supposed to hold while the 'system' is not being 'disturbed' by what could be called an 'observation'. On the other hand, while the 'system' is said to be 'interacting with a measuring apparatus', it is clear that this dynamical language becomes irrelevant, and that an entirely different kind of language is needed. For example, it has been said that the wave function 'collapses' as soon as a human-being has 'knowledge' of the results of this interaction, or that 'potentialities are actualised' when 'the apparatus interacts with the observed system', or that 'the information in the wave function enters the consciousness of an observer' or something else of the sort. What is crucial here is that notions like 'collapse', 'knowledge', 'actualisation of the potentialities', 'information', 'consciousness', 'observer', etc., are *inherently non-dynamical in nature*. That is to say, they simply cannot be expressed at all in the language of dynamics, with its terms like 'field', 'particle', 'position', 'momentum', 'differential operator', 'wave equations' and the like.

Nevertheless, when one starts with a dynamical language for Schrödinger's equation and the wave function, one is tacitly committed to extending this language to cover all aspects of the theory, including what is called the 'measurement process'. Otherwise, it would appear that essential aspects of the total situation are being left out, so that at the very least, the theory must seem to be highly 'incomplete'. The drive for 'completeness' that is inherent in the older dynamical language and its extension to Schrödinger's equation then leads inevitably to the formulation of the 'measurement problem', i.e. 'How is it possible to describe the "measurement process" in some kind of extension of the dynamical language?' This problem would, on the face of it, appear to be insoluble, because the terms of dynamics are so far from those having to do with 'observation' that it is extremely difficult to see how they could even be relevant in this context. This difficulty is reflected in the fact that those who have worked on the 'measurement problem' have been generally not merely unable to agree, but unable even to meet sufficiently for one school to appreciate what the other school is trying to do.

In my view, this kind of unclarity in basic premisses is not the unique responsibility of any individual, or of any particular school. Rather it is the necessary result of the tacit commitment of physics to the language of dynamics with its implied drive towards 'completeness'. Therefore, to be free of this kind of confusion, what is needed is that we begin to develop a different language, that does not commit us to dynamical notions as basic, so that room is left for our thinking to move in qualitative new directions.

In the development of our theory of hidden variables, we shall therefore have to be careful not to 'pour new wine into old bottles' by using current dynamical terms of description and inference. But, of course, it is not possible suddenly to switch from one language to another with a very different general structure. So a transitional 'intermediate' treatment is needed, in which we introduce new terms gradually, while we relate these, whenever possible, to the corresponding older terminology. To facilitate this procedure, we shall always put words referring to the older terminology in quotation marks.

To begin to express this kind of 'hidden variable' theory mathematically,[2] we first define the 'non-hidden' variables which are taken to be the coefficients, c_r, in an orthonormal expansion of the wave function, with the basis χ_r

$$\psi = \sum_r c_r \chi_r.$$

We then introduce a set of 'hidden variables', ξ, which are defined in another Hilbert space, 'dual' to the c_r. That is, we write

$$\xi = \sum_s b_s \xi_s$$

where ξ_s is a set of orthonormal functions 'dual' to the χ_r. It is important to emphasise that in spite of these obvious similarities ψ and ξ are different in key ways. For example, they do not satisfy the same law, or even similar laws. Indeed, as will be seen, in our theory, ψ will tend to reflect the aspect of necessity while ξ will be largely an expression of contingency, so that in the context under discussion ξ does not in general satisfy any well defined laws of motion at all. (Other differences between ψ and ξ will emerge as we go along.)

Our proposal for the 'non-hidden' variable, ψ, is that it shall follow an equation similar to that of Schrödinger's in certain ways, but different from it in other ways. These differences are crucial, in the sense that, (as will be seen), they help to bring basically non-dynamical notions into the foundations of the theory.

In expressing this equation we begin, for the sake of simplicity, by choosing an orthonormal basis for ψ, such that what is currently *called* the 'observable that is being measured' is diagonal. (Remembering, however, that we shall gradually cease to use such terms, as our new language steadily acquires the capacity to 'take over'.) Our equation then takes the form

$$\frac{\partial c_r}{\partial t} = \frac{\hbar}{i} \sum_s H_{rs} c_s + \lambda \sum_s J_s \left(\frac{J_r}{K_r} - \frac{J_s}{K_s} \right)$$

where H_{rs} is the matrix element of the usual Hamiltonian operator, and $J_s = |c_s|^2$, $k_s = |b_s|^2$, while λ is a coefficient, whose meaning will be discussed presently.

Evidently, the first term on the right hand side of the above equation corresponds to the usual Schrödinger theory. The second term, however, is of a very different nature. First of all, it has non-linear dependence on the c_r. Secondly, it depends on the 'hidden variables', ξ_r (also in a non-linear way). And finally, it depends on a new kind of quantity represented by the term, λ.

The details of the behaviour of λ will not, however, be significant at the present stage of the development of the theory. Only certain general features of this behaviour will be relevant for this purpose.

What we are indeed proposing is that λ is a function that depends on the over-all environment of what is currently called 'the system under observation'. In a way, such notions are not entirely unknown in physics. Thus, Mach has proposed, in effect, that the inertial properties of an object might depend on the over-all distribution of very distant stars. Similarly, we are suggesting that not only the inertial properties, but also, other equally 'basic' properties (such as those relative to the coefficient λ) may depend on the general environment. But this environment is no longer being restricted to the distant stars. Rather, it may include *any* feature of the environment, in ways that are open to further specification as the theory is developed.

In particular we are proposing that when it is said in the older language that a 'measurement is taking place', so that 'the measuring instrument interacts for a certain period of time with the observed system', the corresponding description in our language is that environmental conditions are such that λ is large for the period of time in question. λ will then be small again for the period of time in which the 'system' was said in the old language to be 'undisturbed'. It will then be large again for the next period in which the 'system' was said in the old language to be 'interacting with another measuring instrument'. For the time in which λ is negligible Schrödinger's equation will of course be satisfied and this is all that is needed to lead to results that are in agreement with those of the usual theory in the context in which the latter is relevant.

Now, if the content of every description has its ground in the totality of all that we perceive (and all that we know) it is evident that this field is too vast, varied and undefinable even to be described. Indeed, what actually happens is that some aspect of the total is singled out as directly relevant so that it stands out in what may be called the *foreground* of the discourse. All that is not thus singled out (i.e. all that is not directly relevant) then necessarily falls into what may be called the *background*. This is the primitive step behind any act of abstraction (i.e. taking out) of what is essential to the content of a discourse.

Such an *act of abstraction* is evidently a necessary prelude to every discussion, whether in physics or in any other field. However, because of certain fortuitous features of its historical development (mentioned in the previous sections) physics has come to aim towards 'completeness' of description as its ideal. In other words, it is commonly believed that the language of physics ought to be universally relevant in every possible field of discussion, because physics is supposed to give fundamental and generally valid knowledge of 'what each thing is'. But we have seen that if 'basic' properties of each thing may depend on the whole of the general environment (through the coefficient λ in our equations) the attempt to get such knowledge must lead to confusion. Therefore, what is needed is to admit, at the very outset, that physics has no special role of this kind and that it is like every other subject, in that it can deal only in terms of the kind of *inherently incomplete abstractions* to which our notions concerning 'hidden variables' have led.

Let us now return to the discussion of the kind of *description* that is suitable to our particular 'hidden variables' theory. We say that the terms of our equations involving 'hidden' and 'non-hidden' variables are part of the description of the *foreground* but that these terms may well depend on the *background* through the coefficient, λ (as well as in other ways). That is to say, variations of λ in some sense take into account corresponding variations in the background. In order to show how this happens, it will be useful to use the terms of the older language *indicatively* (i.e. merely to 'point to' a certain situation that is generally known, without committing us to the view that these terms give a correct description of the structure of the situation). So we begin with a situation which is indicated in terms of the old language by the phrase 'a process of interaction between the observed system and the observing apparatus is taking place'. But in our language, we say instead that for a specified period of time in which λ is large, *an operation* is taking place, the *result* of which is *revealed* in the foreground, through a certain change in the 'non-hidden' variables, ψ. (The word *operation* has the advantage of having a significant but indirect connection with the mathematical notion of *operator*, a connection whose meaning will show itself as the discussion proceeds.) But in describing such an operation, we shall no longer use the words 'measurement' or 'observation', because these would imply separately existent 'observed systems' and 'observing instruments', which notions have been ruled out as irrelevant and meaningless in terms of our new language.

It will be noted that these views are, in a certain rather significant way, very similar to what is most novel in Bohr's notions. For Bohr also emphasised that physics is in the first instance a *description*, rather than a source of, in principle, complete conceptual images of 'what each thing is'. And in particular he emphasised the unanalysable wholeness of the description of the experimental phenomena by saying that the results of an experiment cannot be separated from the over-all conditions of such an experiment. (This was indeed the basis of his principle of complementarity.) In order to be able to say such a thing, Bohr had to suggest a new way of using language in physics, a way that has not in fact been very generally understood. For example, he frequently indicated that the quantum theory implied that there can be no 'atomic object' which could be 'observed' by a separate 'observer'. And yet, other physicists (including Heisenberg and von Neumann) thought in terms of *two* such systems separated by a 'cut' between them that was, however, 'freely moveable' and therefore in a certain sense ambiguous. It seems significant that Bohr never referred to such a 'cut' at all, and indeed it is clear that it would be inconsistent with his notion of 'unanalysable wholeness' to do so (since such a 'cut' is intended just for the purpose of providing a kind of description that would make possible an analysis of the relationship between 'observer' and 'observed system', considered as separately existent). It should therefore be evident that there are important similarities between the language of our theory of 'hidden variables' and that of Bohr (even though, as will emerge further on, there are also important differences).

3.2. The formal presentation of the theory. From the basic equations for the c_r (given in section 3.1) it follows that

$$\frac{\partial J_r}{\partial t} = \frac{\hbar}{2} \sum_s (c_r^* H_{rs} c_s - c_r H_{rs}^* c_s) + \lambda J_r \sum_s J_s \left(\frac{J_r}{K_r} - \frac{J_s}{K_s} \right)$$

and using the Hermitian character of H_{rs} (i.e. $H_{rs} = H_{sr}^*$), one obtains

$$\frac{\partial}{\partial t} \sum_r J_r = 0, \quad \text{or} \quad \sum_r J_r = \text{constant.}$$

We see then that the 'normalisation' of the wave function is still preserved, independently of time, even though the equation for ψ is non-linear, and even though ψ does not undergo what is usually called a 'unitary transformation'. (Note that for a transformation to be 'unitary' in this usual sense, it is not only necessary that scalar products in Hilbert space be invariant, but also that the transformation coefficients be independent of ψ, a condition not satisfied in our theory).

As already indicated, we shall take the paradigm case to be that in which λ is significantly large only for limited periods of time: in the old language it would be said that 'observed system and apparatus are interacting' in this case. We can further simplify the discussion by assuming λ to be so large for these periods of time that the effects of the Hamiltonian, H_{rs}, can be neglected. In the old language we would be said to be assuming 'impulsive measurements' but we will now say instead that we are assuming *impulsive operations*.

Such an assumption of impulsive operations corresponds closely to the fact that 'quantum transitions' are actually in some sense very abrupt and sudden, requiring much less time, for example, than the calculated period of the initial wave function. Thus in radioactive disintegration of radium nuclei the mean decay time determined by the wave function is over 1,000 years: and yet some α-particles appear immediately (e.g. in less than 10^{-8} s). The introduction of 'hidden variables' makes available a form of description in which it is not only possible but indeed very natural to say that 'quantum transitions' have this sudden and impulsive character, for we are free to choose λ very large, and from this the impulsive character of the operation follows.

For the time in which λ is large, we can then write

$$\frac{\partial c_r}{\partial t} = \lambda c_r \sum_s J_s \left(\frac{J_r}{K_r} - \frac{J_s}{K_s} \right);$$

for this period of time we shall further assume that the 'hidden variables', ξ, do not change appreciably. Therefore we can treat the K_s as constant.

Among all the J_r, there is one, J_t, which satisfies the condition

$$\frac{J_t}{K_t} > \frac{J_s}{K_s}$$

for all $s \neq t$. Evidently, J_t will necessarily increase with the passage of time (because all the terms $(J_t/K_t - J_s/K_s)$ are positive). And, as a simple calculation shows, $J_t/K_t - J_s/K_s$ will also increase, so that J_t must keep on increasing. Since $\sum_r J_r$ is constant, this increase can stop only when all the J_s are zero (for $s \neq t$). Thereafter J_t will remain constant.

So what has been shown is that whichever of the J_r satisfy the condition $J_t/K_t > J_s/K_s$ (or whichever yields a maximum J_s/K_s) will have to increase and to yield $J_s = 0$ for all $s \neq t$, *independently of further details in the initial conditions*. Symbolically, if we represent all possible initial conditions of the c_s by a full circle (fig. 1), then this full circle is divided into sub-regions (labelled by A, B, C, D, etc.) such that all points initially in any given sub-region must arrive at the same final result. This sub-region is determined by the K_s (hence by the 'hidden variables' ξ) as well as by which operator happens to be diagonal in this particular environmental background (or by what would be called the 'experimental conditions' in Bohr's language).

It is clear then that the final result must depend in each case on the 'hidden variables'. We shall assume (as is done in insurance statistics and in statistical mechanics) some kind of *probability measure* for these hidden variables, that will be valid within suitably limited contexts. Such a probability distribution over the 'hidden variables' will of course imply a corresponding probability for the results of a measurement.

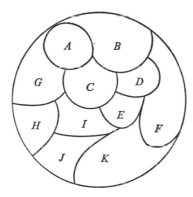

Fig. 1

The only remaining question is to ask whether there exists a normal probability measure that will yield results equivalent to those of the quantum theory, as it is now formulated. The answer is that such a measure does exist. Indeed, if we assume the 'normalisation' condition, $\sum_r J_r = 1$, $\sum_r K_r = 1$, it is easy to show that a uniform distribution of the b_r in their Hilbert space leads to the usual quantum theoretical probabilities for the results of measurements.

It is instructive to illustrate this conclusion in terms of a Hilbert space with only two orthonormal axes (representing, for example, a 'spin' having two possible values in any specified direction). Our equations become

$$\frac{\partial J_1}{\partial t} = \lambda J_1 \left(\frac{J_1}{K_1} - \frac{J_2}{K_2} \right), \quad \frac{\partial J_2}{\partial t} = \lambda J_2 \left(\frac{J_2}{K_2} - \frac{J_1}{K_1} \right)$$

with
$$J_1 + J_2 = 1, \quad K_1 + K_2 = 1.$$

We see that if $J_1/K_1 > J_2/K_2$, then J_1 increases to unity and J_2 drops to zero, while if $J_2/K_2 > J_1/K_1$ then J_2 increases to unity while J_1 drops to zero. This covers all possibilities (except for a set of measure zero which can be left out of consideration).

The normal probability measure is uniform on the hypersurface in Hilbert space $|b_1|^2 + |b_2|^2 = 1$. Writing $b_1 = U_1 + iV_1$, $b_2 = U_2 + iV_2$ we have,

$$U_1^2 + V_1^2 + U_2^2 + V_2^2 = 1.$$

This is a three-dimensional hyperspherical surface in a four-dimensional Euclidean space. The element of volume in this space is $d\Omega = dU_1, dV_1 dU_2 dV_2$. If we use the polar definition of complex variables, $b_1 = r_1 e^{i\phi_1}$, $b_2 = r_2 e^{i\phi_2}$ and with $r_1 = r\cos\frac{1}{2}\theta$, $r_2 = r\sin\frac{1}{2}\theta$, we obtain

$$d\Omega = r^3 dr \sin\theta \, d\theta \, d\phi_1 d\phi_2.$$

Restricting ourselves to $r = 1$, we obtain for the relevant element of 'area'

$$dS = \sin\theta \, d\theta \, d\phi_1 d\phi_2.$$

All that remains to be done is to integrate up the total area that satisfies the relationship

$$\frac{J_1}{K_1} > \frac{J_2}{K_2}, \quad \text{or} \quad \frac{|a_1|^2}{\cos^2\theta} > \frac{|a_2|^2}{\sin^2\theta}.$$

This yields $S_1 = |a_1|^2$. Likewise, the total area satisfying $J_2/K_1 > J_1/K_1$ is $S_2 = |b_2|^2$. Thus, if the distribution of the 'hidden' variables is uniform, we obtain the usual quantum theoretical probabilities of results $P_1 = |a_1|^2$ and $P_2 = |b_2|^2$. (These conclusions are easily extended to the case of a general Hilbert space, to yield $P_n = |c_n|^2$ for the probability of a result corresponding to the n-th unit vector in Hilbert space.)

All this holds for the normal uniform distribution. For any other non-normal (hence non-uniform) distribution, different results will be obtained. In particular, for a delta function distribution of the hidden variables, the result is completely determined, according to whether

$$\frac{J_1^1}{K_1} > \frac{J_2}{K_2}, \quad \text{or} \quad \frac{J_2^1}{K_2} > \frac{J_1}{K_1},$$

for the particular values of the hidden variables corresponding to the 'delta function'.

It is evident that since our basic equations are non-linear, it is quite natural that the conditions of von Neumann's linearity postulate for quantum theoretical averages need not be valid for general distributions of the 'hidden variables' (though, of course, they are valid for the normal uniform distribution).

A question that arises naturally at this point is whether it is possible actually to obtain non-normal distributions that would give rise to new kinds of experimental results going outside the whole framework of the quantum theory, in its current general formulation. This possibility depends on the behaviour of the 'hidden variables', b_s. Of course this behaviour is necessarily unknown, for the present. But we may find it instructive to consider what is implied by various possible assumptions on this point.

The simplest assumption is that after the *impulsive operation* is over (so that in the language of the current quantum theory, the 'measurement process' is said to be finished), the hidden variables simply remain constant.

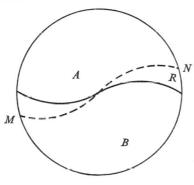

Fig. 2

Consider for example the case of what has been called 'measuring the spin of an atom in the z-direction'. In our language, we say instead that a z-spin operation has taken place. Let us suppose that such an operation gave a result corresponding to a 'spin' of $\hbar/2$, so that $|c_1|^2 = 1$, $|c_2|^2 = 0$, and the 'hidden variables' were described statistically by a uniform normal distribution. We further suppose that a corresponding spin operation for some direction other than z then takes place. We now let a circle symbolically represent all the possibilities for the *hidden variables* (and not for the c_r, as it did in fig. 1). If we denote quantities referring to the new direction by primes, this circle will be divided into two parts, A and B, representing states of the hidden variables that respectively satisfy

$$\frac{J_1'}{K_1'} > \frac{J_2'}{K_2'}, \quad \text{or} \quad \frac{J_2'}{K_2'} > \frac{J_1'}{K_1'} \quad \text{(see fig. 2)}.$$

If, in fact, the result corresponds to ψ_1', then we can deduce that the hidden variables are in the region corresponding to A, and not in that corresponding to B.

It follows then that if a further spin operation takes place in yet another direction, the probabilities will be different from those given by the quantum theory in its current formulations; (this situation is indicated by the dotted line MN). It is clear that for example the probability of a result corresponding to a 'spin' of $-\hbar/2$ in this new direction would be proportional to the intersections between the area of A and the corresponding area on the right hand side of the line MN, to yield a net result, denoted by R, which would have a very different statistical weight from that implied in the formulae given by the current quantum theory.

It should be noticed, nevertheless, that if the third spin operation corresponds to the same direction as the second, then the condition $J_1''/K_1'' > J_2''/K_2''$ continues to be satisfied (because $J_2' = 0$ after the second operation). So any particular operation will be reproducible (as is also assumed in the current quantum theory). Indeed, even if the 'hidden variables' should change between operations, the condition $J_1''/K_1'' > J_2''/K_2''$ would still be satisfied. So this feature of reproducibility is essentially independent of detailed assumptions about the behaviour of hidden variables.

More generally, however, one sees that in a series of spin operations, corresponding to what has been called 'measurements of incompatible observables', the 'hidden variables' can be progressively limited and defined. In this process, they cease to be 'hidden' since they can now reveal themselves in a *new kind* of statistical distribution of results, not obtainable in terms of the current quantum theory.

It has thus emerged that the distinction between 'hidden' and 'nonhidden' variables is not an absolute and final one, but that it can vary in accordance with the context. So this distinction has at most a rather limited kind of relevance. Therefore, it would be best to cease to regard this distinction as a basic in our description from now on. Indeed, we shall introduce the new terms, the *contingent part of the description* (represented for short by the *contingent variables*) and the *non-contingent part* (represented by the *non-contingent variables*). One must remember of course that what was contingent at one stage in the development of a theory can become non-contingent, or necessary, at another.

If the contingent parts of the description (i.e. the 'hidden variables') can reveal themselves in the way described above through non-normal distributions, one may ask why they have not yet done so in experiments that have thus far been carried out. A very simple answer is that the contingent variables may be subject to some kind of stochastic process. If this process defines a certain characteristic relaxation time, τ, then successive operations separated by an interval greater than τ would be described by the normal uniform distribution of the contingent variables, leading to the usual quantum theoretical probabilities. But operations following each other in times shorter than τ would reveal new statistical results. And we may assume that τ is shorter than any time interval that has been relevant until now in physical operations, thus explaining why the contingent variables have not yet revealed themselves.

Naturally, the value of the parameter τ is at present almost entirely unknown (other than through the limit that it is shorter than present 'measurements' allow). One may speculate that it is perhaps related to another parameter with the dimensions of time which also has to do with the statistical properties of matter in bulk and with the quantum. That is $\hbar/kT$, where T is the temperature. It is unclear, of course, whether the relevant temperature is that of the apparatus or of the observed systems (or of both, in some combination). Such a speculation may nevertheless be useful if it helps to suggest experiments that could test these notions. [3]

It is important however not to commit ourselves prematurely as to the properties of the b_s. Indeed, as we have already seen, they are in the first instance mainly a further addition to our *language*, enabling us to describe certain kinds of contingency. To emphasise the full potentialities of this notion we may note that the b_s need not represent simple stochastic variables, belonging to what has been called the 'observed system' in the older language. Rather, they may constitute a new kind of description that takes into account various features outside the scope of the non-contingent variables, c_s. In the older language, this would mean roughly that the b_s 'reflect' some aspect of what has been called the 'observing apparatus' or even 'the general environment' and not merely what has been called the 'observed system'.

As an illustration of this kind of 'reflection' of contingent variables consider, for example, the image charge of an electron facing a metal surface. In a rough treatment, this image charge may be assumed to have a well-defined position, in relationship to that of the electron. But because of the atomic structure of the metal, as well as because of all sorts of irregularities in the surface and in the crystal structure, the image is not actually completely defined in position by the electron. Rather it is subject to 'fluctuations' that are contingent on the fortuitous details of the atomic and crystal structure. In certain contexts, one might define a statistical measure of these contingent variables in terms of some kind of probability distribution of image positions. Similarly, the b_s may represent a set of contingent variables corresponding to fortuitous variations in the total context of the description of a kind that is as yet unknown.

To assume in this way that the equation for the non-contingent variables c_s depend on a statistically distributed set of contingent variables b_s, is evidently in itself a significant step outside the general descriptive language of dynamics. In this latter language, it is indeed taken for granted (as has already been pointed out) that a dynamical system can be analysed into its constituent parts or elements and that the equations governing the changes of these elements with time are simple necessities independent of all contingency. To be sure, these elements interact, but the laws of interaction are specified entirely in terms of the dynamical properties of the elements and do not depend fortuitously in any way at all on contingency. Indeed, in dynamics, all contingency is restricted as a matter of course, to the initial conditions of the constituent elements. Thus, to have the laws of change and development themselves depend on a set of statistically distributed contingent variables is a step outside the whole framework of dynamics. This means indeed that each aspect abstracted in the field of discussion has a kind of *inherent incompleteness* (in a sense that is over and above that corresponding to the dependence of the coefficient λ on the general background).

There is still another kind of incompleteness that is very relevant to the language structure of our theory. To see this let us go back to our basic equations (for the special case of an *impulsive operation*):

$$\frac{\partial c_r}{\partial t} = \lambda c_r \sum_s J_s \left(\frac{J_r}{K_r} - \frac{J_s}{K_s} \right).$$

Because $J_r = |c_r|^2$ and $K_r = |b_r|^2$, it follows that the form of the equations is not invariant under changes of representation of the wave function in Hilbert space. This lack of invariance is however only a result of the rather special form in which the equations were first put for the sake of simplicity of expression. Indeed, if we now drop this restricted form, we will be able to see further into the full meaning of these equations.

To do this we recall that the equations are now in a representation in which what has in the older language been called 'the operator that is being observed' is diagonal. But in principle every such operator can be re-expressed in terms of a set of projection operators, P_r, each of which projects out all eigenvectors of the operator in question except the r-th. We can then write $J_r = \psi^x P_r \psi$ and $K_r = \xi^x P_r \xi$ with $\psi = \sum_r c_r \psi_r$, $\chi = \sum_r c_r \chi_r$, our equations can be rewritten as:

$$\frac{\partial \psi}{\partial t} = \lambda \sum_{r,s} J_s \left(\frac{J_r}{K_r} - \frac{J_s}{K_s} \right) P_r \psi.$$

If we change the representation by writing $\psi = U\psi'$, $P_r = UP'_r U^{-1}$, then it is evident that the form of the equation is now invariant to such a change of basis in Hilbert space.

However, what is new is that a certain set of operators, P_r, now enters the basic equation for ψ. In the language of Bohr, this set is determined by the 'total experimental conditions' (e.g. as to whether 'position' or 'momentum' or some other 'observable' is being 'measured'). But in our language we say that the operators, P_r, depend on the *background* which is of course contingent in the context of our theory. Different kinds of contingent backgrounds then lead to different (and generally non-commuting) sets of such operators. So (as also holds in a way in Bohr's language) we say that the mutual incompatibility of different contingent backgrounds corresponds to the feature of non-commutativity in certain terms of the mathematical algorithm.

We are not however necessarily committed to this particular meaning of the operators P_r, i.e. that they take into account what Bohr has called the 'experimental conditions'. We can extend the meaning by ceasing to regard as basically relevant the difference between a situation called an 'experiment' and any other situation that might arise, either naturally and 'of its own accord' or artificially and through the participation of people who are called 'physicists'. In other words, the P_r can now take into account all sorts of features of the background. Along with the coefficient, λ, they imply that the very equations that describe the behaviour of the foreground are crucially dependent on the background, in ways that are open to further specification as the theory is developed.

We see that there is yet another kind of *inherent incompleteness* in the theory. For the basic equations depend not only on statistically distributed content variables such as the b_s, but also on *non-statistical contingent variables*, which are described by the P_r. Indeed, these latter take into account what are in general fortuituous features of the overall environment and background that have to be described in terms of new contexts, beyond the one that is now under discussion.

REFERENCES

(1) von Neumann, J. *The Mathematical Foundations of Quantum Mechanics* (Princeton, N.J.: Princeton University Press, 1955).
(2) Bohm, D. and Bub, J. *Rev. mod. Phys.* (1966) **38**, 453.
(3) Papaliolios, C. *Phys. Rev. Lett.* (1967) **18**, 622.

BEYOND WHAT?

C. W. KILMISTER

The point of view which I want to take in this paper is that we need to ask, first, what is it exactly that quantum theory has achieved, and then to ask in what way one needs to go beyond it—not in the sense of applying it to wider fields (e.g. biology)—but what needs to be done to formulate, and therefore to understand, the theory better. Essentially the view I am taking can be summarised by saying that the nature of a scientific explanation is at issue here. It is true that there have been times when quantum mechanics has provided, or come near to providing, what I would regard as a satisfactory explanation, but I want to argue that these occasions are much fewer than supposed. By saying what it *has* provided at each stage I hope both to make clearer my idea of what constitutes a satisfactory scientific explanation and also to see what more is needed in the case of quantum theory.

One can see in a general way the drift of the argument by considering the explanation[1] which has been seriously advanced, whether correctly or not, for the alleged fact that raindrops in a storm have certain discrete masses:

> ...as Wilding Köller and V. Byerknes tell us. The drops are of graded sizes, *each twice as big as another* beginning with the minute and uniform droplets of an impalpable mist. They rotate as they fall and if two rotate in contrary directions they draw together and presently coalesce: but this only happens when two drops are falling side by side, and since the rate of fall depends on the size it always is a pair of co-equal drops which so meet...

In this explanation the existence and magnitudes of integral-valued variables are deduced from the underlying continuous theory. A desire to do exactly this has motivated some of the proponents of 'hidden variables' in quantum mechanics: I do not think it is necessary to follow their example. My purpose is rather to abstract from the theory in this example those properties which make it a satisfactory explanation, in order to enquire later in what other ways a similar result may be achieved.

In such an investigation there is a certain amount of difficulty in knowing what exactly constitutes quantum theory. Since the beginning of the century quantum theory has been the name given to a number of different theoretical systems, all concerned with the explanation of atomic phenomena. I would use the name for all of these systems so that, for example, if I were asked whether to include the symmetry theory (SU_2 or SU_3 groups) as quantum mechanics I would answer 'yes' because physicists have regarded them as applicable to elementary particles. I shall arrange the investigation according to the historical development of the theory. Planck[2] made certain assumptions for deriving the spectrum of black-body radiation and relating the results to thermodynamics, but these assumptions by themselves could hardly be considered a theory, since the essence of a theory is that it should be applicable in *more* than one instance. And although it is often stated that Einstein [3] applied Planck's assumption to the photo-electric effect, this is not really the case. In the paper cited Einstein introduced the idea of the light quantum with the object of reconciling classical Maxwell theory with black-body radiation. We can really count the beginning of quantum theory when this assumption of Planck's together with that of the existence of preferred values of angular momentum were used in the Bohr atom. [3] What did the Bohr atom do? It recognised the existence of integer-valued variables (as indeed it had to do because of the Balmer formula!), but it correlated this with the postulate of the integral character of certain quantities in Newtonian mechanics, i.e. energy and angular momentum.

Bohr's approach is instructive; his starting-point is Rutherford's model of the atom derived by α-ray scattering. [4] He proposes to use this to formulate a theory of atomic structure on the basis of two assumptions motivated by Planck's theory: (i) that there exist stationary states in which the behaviour of the atom may be described classically; (ii) in jumps between these states the classical description is inadequate, but the energy change is correlated by Planck's law with frequency emitted. But these assumptions he first augments with a special assumption that the frequency ν of the radiation emitted during the process of binding an electron from infinity into a stationary state is connected with the frequency ω of revolution of the electron in its orbit; and since at the beginning of the process the electron is at rest, he assumes $\nu = \frac{1}{2}\omega$. Later, when he has shown that the spectrum of the hydrogen atom is accounted for, he says 'while there obviously can be no question of a mechanical foundation of the calculations given in this paper...' it is possible to give an *interpretation*. There then follows the statement that $\nu = \frac{1}{2}\omega$ and Planck's law taken together are equivalent to quantising angular momentum. In *some* ways Bohr comes nearer to a satisfactory explanation here than his successors were to do. The classical description is available most of the time; and for the rest special assumptions are employed which, it is true, are not explained, but which can be expressed in terms of classical concepts by saying that energy and angular momentum are quantised.

Of course this created a formal inconsistency, but it was not this fact that was worrying. The problem was to know what were the right integral relations for more complex problems. For the alkali atoms, for example, it was straightforward to argue that the potential of the nucleus on an electron was modified by the screening effect of the other electrons. This led to a corresponding formula:

$$\nu = Rc \left[\frac{1}{(n_1 - \alpha_1)^2} - \frac{1}{(n_2 - \alpha_2)^2} \right],$$

where now α_1, α_2 were two *non*-integral parameters having different values for the different series. Indeed, the normal Zeeman effect for simplets was also explicable but only so long as an additional selection rule was applied, requiring one quantum number to change by not more than unity. This need for a new rule was a first indication of the kind of difficulty in which the theory was to founder; because the mechanics had not given rise to the integral relations, but had them imposed from outside, there was no way of finding new relations when required. And so the doublet structure of the alkali spectra, which indicated twice as many states as the Bohr theory predicted, and the anomolous Zeeman effect (and the related Paschen–Back effect in which the anomolous Zeeman separation of a normal multiplet in a weak magnetic field goes over into the normal separation of a triplet for strong fields) were beyond the theory to explain. It has indeed been held by Landé [5] that quantum theory, in the sense of the Bohr atom, could have done much more than it did if it had paid due attention to Duane's [6] results. Without taking this view too seriously it is certainly true that the difficulties into which the old quantum theory ran were *partly* human difficulties which would have been surmounted given time, if the new quantum theory had not taken its place. Both the investigation of adiabatic invariants (see Burgers [7] and Ehrenfest [8]), and the much later geometrical development of Synge, [9] improved the technique for applying the theory, but still left the integral relations imposed from outside.

Considering now the difficulties which actually led to the development of the new quantum theory, the principal ones were the complex problem of the multiplets of the alkali and alkaline earth spectra, and their Zeeman separations. The empirical g-formula of Landé had enabled these Zeeman separations to be worked out, but only by the use of half integers, whereas the old quantum theory had always used integers. Moreover the spectrum of neutral helium which had been worked out by the old theory was completely at variance with all the experimental results. But the problem that led most directly to the formulation of the new mechanics arose in the work of Kramers and Heisenberg [10] on dispersion. Heisenberg, indeed, with the dispersion theory, made a real advance in the sense that he suggested confining attention to observable quantities only. This led, in his hands and those of Born and Jordan, to the development of matrix mechanics, [11] which at the very least removed the inconsistencies which had been present from the beginning in the old quantum theory because of the discreteness postulated there for quantities which classically were continuous. Certainly there were great advantages in this; the computational technique which, before 1925, it was not at all easy to know how to apply to a new problem was replaced by an almost complete technique in which the integral properties of variables followed from the formalism. The technique was not quite complete, because in addition to postulating a Hamiltonian $\hat{H}(\hat{p}, \hat{q})$ of the same form as the classical Hamiltonian $H(p, q)$ when p, q are replaced by the operators $\hat{p}$, $\hat{q}$, and also postulating the Hamiltonian form of the equations of motion in the Poisson bracket form:

$$i\hbar \frac{d\hat{q}}{dt} = \hat{q}\hat{H} - \hat{H}\hat{q}, \quad i\hbar \frac{d\hat{p}}{dt} = \hat{p}\hat{H} - \hat{H}\hat{p}$$

it was necessary to determine the commutation rules so that the classical equations of motion

$$i\hbar \frac{dq}{dt} = \frac{i\hbar}{m} P, \quad i\hbar \frac{dp}{dt} = - i\hbar m\omega^2 q$$

were also true.

Now although (to take the example of the harmonic oscillator) one way of doing this is to argue that

$$\hat{q}\hat{H} - \hat{H}\hat{q} = \frac{1}{2m} (\hat{q}\hat{p}^2 - \hat{p}^2\hat{q})$$

$$= \frac{1}{2m} [(\hat{q}\hat{p} - \hat{p}\hat{q})\hat{p} + \hat{p}(\hat{q}\hat{p} - \hat{p}\hat{q})]$$

$$= i\hbar\hat{p}/m,$$

which readily proves that $[\hat{q}, \hat{p}] - i\hbar$ anti-commutes with $\hat{p}$ and similarly for $\hat{q}$, it is equally possible to argue that

$$\hat{q}\hat{H} - \hat{H}\hat{q} = \frac{1}{2m}[(\hat{q}\hat{p} + \hat{p}\hat{q})\hat{p} - \hat{p}(\hat{q}\hat{p} + \hat{p}\hat{q})],$$

so that $(\hat{q}\hat{p} + \hat{p}\hat{q}) - i\hbar$ must commute with both $\hat{p}$ and $\hat{q}$. The first case strongly suggests the usual commutation rule $[\hat{q}, \hat{p}] = i\hbar$, whereas the second (equally likely) one suggests instead $\hat{q}\hat{p} + \hat{p}\hat{q} = i\hbar$, which leads [12] to a second kind of harmonic oscillator with two energy states only.

In what way, then, was matrix mechanics deficient as an explanation? Apart from the ambiguities just mentioned, it was at the opposite pole to the old quantum theory. There 'position' and 'momentum' and the other dynamical variables referred to the classical concepts but under the additional constraints of quantisation. Here the same names were given to matrices (or linear operators) which had the integral properties but seemed to bear no resemblance to the classical concepts. The rôle of the classical picture was merely to serve as a *naming procedure*; that is, when the operator

$$\hat{H} = \frac{1}{2m}\hat{p}^2 + \tfrac{1}{2}m\omega^2\hat{q}^2$$

was defined, the corresponding classical expression was used to give the *name* 'harmonic oscillator' to the system. And when a set of quantum entities were 'prepared with momentum in the z-direction', it was the operators $\hat{p}_x$, $\hat{p}_y$ which were set equal to zero, because of the classical analogue. But apart from this naming procedure, the connection with continuous quantities had vanished; so that the original problem that the integral relations had to be imposed arbitrarily instead of growing out of the dynamics was replaced by the opposite one. That is, the integral relations were almost completely incorporated, but into a formal system which, although it was called a system of dynamics, was one only by analogy with the corresponding classical expressions.

From one point of view the techniques introduced the following year by Schrödinger were only a particular representation of the matrix operators in a form very useful for calculating, and with a particular choice of the commutation relations. But with the Schrödinger point of view the wave function entered, and this has been the cause both of a great deal of misunderstanding of quantum mechanics and also of its nearest approach to a real explanatory success in the sense in which I am using the phrase. The misunderstandings have arisen from the way that the wave function has been used to get nearer to an explanation known by the name of the Born interpretation (some sort of probability interpretation at least of charged particles occurs already in Schrödinger's paper and Born [13] in 1926 proposed less than the full scale Born interpretation as now understood). If one wishes to take the Born interpretation seriously one must surmount a formidable number of difficulties. Some of these are the difficulties of any probability theory, sharpened by the use made of the theory here. But in addition there are others, for example the fact noted by Suppes, [14] that the joint probability distribution of position and momentum even for simple systems may be determined by a characteristic function which does not correspond to a proper probability distribution.

Perhaps these difficulties are worth surmounting because some sort of probability interpretation is involved in understanding the use of the wave function for any of the numerous applications of the theory which involve the tunnel effect. Specifically there is very good experimental agreement about the behaviour of a stream of particles encountering a potential barrier whose height is respectively either greater than or less than the incident energy. Here the Schrödinger technique gives a discussion which, taken with the corresponding phenomena in optics, gives a satisfactory intuitive explanation. To summarise the position reached here, the theory in 1926 had come near to being a satisfactory explanation but the trouble was that it was still not complete. The particles involved in experiments like the tunnel effect do not always behave like classical particles so that the classical model is not always there to help, but both the Heisenberg and Schrödinger formalisms require this classical background to complete their formal structure. A new concept of particle is needed and is not provided by the theory.

The possibility remains that the complete explanation might be possible on these lines, although no one has been successful in finding it. I think the strongest evidence against this is that, if these were the right lines, it surely ought to have been only a small matter to modify the theory so that it was consistent with special relativity. In fact, however, the small modification proposed by Dirac in 1928, although it led to one of the really great explanatory triumphs of the theory, also led to the realisation that the theory was not, as expected, describing single particles at all but needed a second quantisation.

I refer here, first, to the explanation of half-integral spin. Pauli,[15] in his paper on the alkali spectra, had suggested that the doublets were due to a new property of the electron, 'a two-valuedness not describable classically'. This explanation was carried a step farther by Uhlenbeck and Goudsmit [16] by postulating an intrinsic spin for the electron. Dirac's investigation [17] seemed to show that the electron spin was merely something that arose from Lorentz invariance, and this was a very great advance. At the same time a closer inspection of Dirac's argument leads to some doubts.

These doubts arise from a study of the details of the argument given for electron spin by Dirac. Essentially, the non-commuting of a component of orbital angular momentum with the Hamiltonian (corresponding to its non-zero role of change) is *balanced out* by a corresponding non-commutation of the component of the matrix σ (which arises simply from the algebra of the finite matrices employed in factorising the Hamiltonian). The two quantities balancing out are of different kinds; perhaps this may not matter, but it makes the calculation, as an explanation, less than convincing.

In any case the hole theory then showed, first, that a completely new interpretation was required and, secondly, brought in the new computational techniques of quantum electrodynamics. This subsumed the earlier theory and allowed more refined calculations, and in its early stages had considerable success. It is perhaps desirable to mention in detail the Lamb–Retherford experiment. According to Dirac's theory the hydrogen atom should have degenerate energy levels, the $2p_{\frac{1}{2}}$ and $2S_{\frac{1}{2}}$ levels having the same energy. The Lamb–Retherford experiments showed these two levels to differ slightly in energy, the $2S_{\frac{1}{2}}$ lying about 3 microelectronvolts above the $2p_{\frac{1}{2}}$. The experiment was actually carried out by producing an atomic beam of hydrogen from an oven and exciting some of the atoms, by a crossing beam, to the meta-stable $2S_{\frac{1}{2}}$ state. After the atoms were aligned by passing them through a uniform field, they were detected by measuring the current ejected from a metal electrode on which they impinged. A radio-frequency electric field was applied to the atoms and the variation of the current measured as a function of frequency. There is a resonance at the frequency effective in producing transitions from $2S_{\frac{1}{2}}$ states to $2p_{\frac{1}{2}}$ states, and from this resonance frequency the energy separation can be obtained. This experiment is described at some length to show how successful the theory is in dealing with the crude facts of experiment, though, to be sure, the feeling of elation at a prediction correct to 10^{-6} of an atomic unit is somewhat marred by studying the original calculation of Bethe.[18] He included the contribution of the electromagnetic proper field to the electron mass (although this contribution is infinite) and so calculated the Lamb shift as the difference between the contributions of the proper field (a) in the free state, (b) in the Coulomb field of the proton on a non-relativistic basis. The reformulation of all this in a fully co-variant way has, it is true, been carried out since, but the necessity of mass renormalisation remains.

Nonetheless such considerable success for quantum electro-dynamics has made some people, for example Dyson,[19] take a view about the theory which is diametrically opposed to our requirement of an explanation. Dyson insists that such a successful theory must be in itself an explanation, despite its lack of intuitive clarity. Instead of requiring more by way of explanation, such critics simply say that the process of learning quantum theory is a different kind of process from, for example, that of learning Newtonian mechanics, indeed that it is a process in which the pupil spends a period of some months of acute discomfort, asking various questions about the meaning of what he is doing and being unable to find the answers to them, until at the end of this period of initiation he understands that these are the sort of questions that it is not permissible to ask in quantum mechanics. When he has understood this, so these critics say, he will have under-stood the theory. I cannot accept this idea of understanding because it gives no assurance that we shall know how to apply the concepts in a new situation which may develop in the future. Moreover since quantum field theory outside of electrodynamics has achieved little in the way of quantitative agreement in the high energy field it seems as if the view is over optimistic. A clue to a new approach, which I suggest needs to be followed up, is contained in the various methods called generally the theory of the analytically-continued S-matrix.[20] Here there is a systematic attempt not to appeal to the space-time continuum as far as position is concerned and this draws attention to the fact that all the earlier versions of quantum mechanics have begun with a continuous formalism (classical mechanics) in Euclidean space and time, and have provided techniques for imposing integral-valued variables on this continuum. Some discussions have taken place about the role of time but about space nothing has been said, and the assumption has always been, at least locally, a Euclidean space at every instant. This is suspicious because this space was an artificial construct of the Greek geometers and after such a long use merits another look. All physical measurements are rational numbers and the device invented by the Greeks in order to construct Euclidean space is the real number field. This device has, it is true, a respectable history, dating back as it does to the discovery by Pythagoras that $\sqrt{2}$ was not a rational number, and the subsequent repair by Eudoxus of the damage done to geometry. The needs of the calculus caused the rediscovery of the construction of the real number field in the nineteenth century. It is worth while dwelling for a little on exactly what was achieved. The rational numbers form a field which is totally ordered by magnitude, and this ordering satisfies the axiom of Archimedes. The achievement of the mathematicians was to construct, by adjoining to this field the limits of convergent sequences, a larger totally ordered Archimedean field, indeed the largest one possible. The cleverness of this achievement has hypnotised everybody since. Even when Dirac introduced the δ-function it was treated as an extension of the idea of a function rather than as pointing to inadequacies in the field of numbers. Similarly, divergences in field theories were treated pathologically. All this points to the need for a radical reconstruction.

Now such a reconstruction has recently been carried out in the form of non-standard analysis. [21] Without entering into details, a new field is constructed containing the real numbers as a sub-field and the new field is again totally ordered. But of course this ordering can no longer satisfy the axiom of Archimedes, since the real number field was the largest one with that property. The new field contains both infinitesimals and infinite numbers. The finite members of the field can be written as a sum of a real number and an infinitesimal. It may indeed be the case that such a field has a great deal to offer to quantum mechanics in discussing the divergences but the lesson which I want to draw here is that all the techniques described above fail to be explanations or to lead to explanations because they try to relate the integers to the real number field. It may be that this attempt was doomed to failure because the real number field was an artefact with no prior claim to consideration.

We are now able to see more clearly what *would* constitute an explanation here. An explanation will have to make it clear that in some sense there are different kinds of physical quantities, and show how the important problem of the nature of this difference can be further investigated. Up to now the difference has been thought of as that between the set of integers and the continuum, i.e. between discrete and continuous magnitudes. In order to tackle this problem without first assuming a particular form for the 'continuous' magnitudes

—since the real number field no longer holds its unique privileged position—it is essential to begin at the other end. That is, following the lead given by Heisenberg in 1925, the theory should begin with the observed quantities (the integral values characteristic of quantum theory) and construct the other types of physical variables (like mass) by means of the relations between integral measurements. In this way the experimental results can be used to form a foundation for whatever substitute is appropriate for the continuum, instead of being fitted into a pre-determined frame. Once the mass-like quantities can be defined in this way by means of combinatorial relations, rather than fed into the theory as arbitrarily imposed ratios, there is in principle an explanatory theory.

What advantages should such a theory give? Two possibilities may be suggested, although at this stage it can only be conjecture.

(i) At present a major gulf between combinatorial and analytic methods is evidenced by the different approaches of symmetry theories (SU_3 etc.) and the analytically-continued S-matrix. The explanatory theory suggested could scarcely be worth the trouble of elaborating if it could not make some major contribution to bridging this gulf.

(ii) There have been attempts (e.g. Birkhoff and von Neumann,[22] Piron[23]) to construct lattice-theories to serve as a convenient axiomatisation of quantum theory. A problem here is, what particular axioms are to be adopted for the lattice? Those chosen are chosen with an eye on the consistency between the relations of lattice elements and the relations between subsets of a Hilbert space. Yet the Hilbert space characterisation itself is beset with many difficulties; so that an explanatory theory ought to be able to construct the lattices in an explicit manner, and so settle the question of axiomatisation.

REFERENCES

(1) Thompson, D'Arcy. *On Growth and Form* (Cambridge University Press, 1942), chap. 2.
(2) Planck, M. *Verh. dt. phys. Ges.* (1900) **2**, 202, 237.
(3) Bohr, N. *Phil. Mag.* (1913) **26**, 1.
(4) Rutherford, E. *Phil. Mag.* (1911) **21**, 669.
(5) Landé, A. In various writings, especially: *New Foundations of Quantum Mechanics* (Cambridge University Press, 1965).
(6) Duane, W. *Proc. natn. Acad. Sci. U.S.A.* (1923) **9**, 158.
(7) Burgers, J. M. *Annln. Phys.* (1917) **52**, 195.

(8) Ehrenfest, P. *Naturwissenschaften* (1923) **11**, 543.

(9) Synge, J. L. *Geometrical Mechanics and de Broglie Waves* (Cambridge University Press, 1954).

(10) Kramers, H. A. and Heisenberg, W. *Z. Phys.* (1925) **31**, 681

(11) Born, M. and Jordan, P. *Z. Phys.* (1925) **34**, 858.
Born, M., Jordan, P. and Heisenberg, W. *Z. Phys.* (1925) **35**, 557.

(12) Hund, F. *Materie als Feld* (Berlin: Springer, 1954).

(13) Born, M. *Z. Phys.* (1926) **38**, 803.

(14) Suppes, P. *Philosophy Sci.* (1961) **28**, 378.

(15) Pauli, W. *Z. Phys.* (1925) **31**, 373.

(16) Uhlenbeck, G. E. and Goudsmit, S. A. *Nature* (1926) **117**, 264.

(17) Dirac, P.A.M. *Proc. R. Soc.* (A) (1928) **117**, 610.

(18) Bethe, H. A. *Phys. Rev.* (1947) **72**, 339.

(19) Dyson, F. J. *Scient. Am.* (September 1958).

(20) See, for example, G. F. Chew in:
Jacob, M. and Chew, G. F. *Strong Interaction Physics* (New York: Benjamin, 1964).

(21) Schmieden, C. and Laugwitz, D. *Math. Z.* (1958) **69**, 1.
Robinson, A. *Non-standard analysis* (North-Holland, 1966).

(22) Birkhoff, G. and von Neumann, J. *Ann. Math.* (1936) **37**, 823.

(23) Piron, C. *Helv. phys. Acta* (1964) **37**, 439.

DISCUSSION: SPACE-TIME ORDER
WITHIN EXISTING QUANTUM THEORY

(VON WEIZSÄCKER): I wonder if I could put Bohm's view like this:
In general we start out with a space-time structure to which we
add quantum theory, and we describe objects by relying on the
theory to tell us what is going on in terms of the space-time
structure. However, you say that the way in which we generally
look at space-time structure may be too narrow and that quantum
theory itself teaches us about an order which is somehow connected with what we originally knew as space-time structure, but
which is richer.

(BOHM): Yes.

(VON WEIZSÄCKER): So far, I think, I follow you. On the other hand
one might say that since we are here learning something about
space-time structure from quantum theory, there is no reason why
we should not deduce the whole of what we call space-time structure from quantum theory.

(BOHM): I must object to that. What does the expression 'quantum
theory' mean? It represents a certain union of formal and informal language. We are discussing the formal language of
quantum theory, more or less as Aharonov says in his paper, using
it as a clue to change the ordering of the informal language.

(AHARONOV): What I said was that we have not fully appreciated the
content of quantum theory because of preoccupation with the
negative aspects of it—what limitations we have, and so on. We
are not asking honestly and with sympathy, what the new experimental set-ups are that quantum theory dictates as contrasted
with the classical set-ups.

I am not denying that each individual part of the apparatus is
classical. What I do say is that the arrangement of the parts makes
sense only on the basis of quantum mechanics. These arrangements
would look meaningless without quantum mechanics.

This fact has mostly been overlooked. Arrangements that Bohr,
for example, considered were all classical arrangements in the sense
that they were either arrangements for measurement of classical
waves or arrangements for measurement of classical particles.

There were no essentially new arrangements in his discussions, and that is what I consider missing. There are in fact new arrangements that quantum theory tells us about, like the relative rotation by 2π, which is not a classical arrangement because it rotates the spin vector by π. Classically, nobody would consider relative rotation by 2π, they would say it is nonsense to do so.

(BOHM): The difficulty is, the language you are using to try to explain things is in fact getting them a bit mixed up. You are tacitly suggesting that we consider an order in the experimental situation which is different from the classical order of events. It is not that we deny classical physics, but that for certain purposes it is not relevant, and that things you have called non-classical arrangements, which are really *new* orders, are relevant. Actually it doesn't matter whether they are classical or quantum mechanical or anything else; what you have to do is watch for the relevance of new orders.

You are suggesting, quite rightly, that even for the fundamental purposes of physics, orders which are as different from classical physics as the order of the flowers is from classical physics, may be relevant because you are looking at rather complex arrangements and saying 'this is what we observe and not momentum or position'.

First of all however, I think the language should change. What you are suggesting requires a different language of experimentation which will then lead the experimenter to look for something different. The theorist does not deal directly with experiment, he deals mostly with language.

(BASTIN): What did you mean by another 'order'?

(BOHM): It is rather difficult to explain, and would take a long time.

(AHARONOV): Except by example: this rotation by 2π is something you would not consider relevant in classical theory.

(BASTIN): And you call it another 'order'?

(BOHM): The order becomes important when you think about the apparatus. Ordinarily you say that if anything rotates by 2π then the change in the order is irrelevant. But now if we say that there is some relevance in the difference between 2π and 4π, we have another notion of order, of objects, or whatever you wish. We don't care if this is classical or not, if the difference in the 2π case is physically relevant it is another order.

(VON WEIZSÄCKER): What you have been using here is essentially the spinor mathematics, and I think one might be able to build up the concepts of objects and of space and time together (rather as was discussed by Penrose†) in the following way:

You say that all objects, especially particles and all that consists of particles, are really built up of primitive objects which are not spinor functions in space where you already have space to begin with, but which are spinors which are not functions of anything. The multiplicity which you have in space is then brought in by having many objects of this sort, and having them going together. I think I would be able to show that this would lead to a space which has all the properties you are describing because it would essentially be a spinor space and not a vector space. In this sense what you have been saying here, looks to me like an example of a thesis which I would be prepared to defend, namely that space-time structure as we know it can be derived by starting out with a principle of simple alternatives in quantum theory, i.e. by spinors which are not functions of anything. If then we add together as many of these objects as we like what we shall get will be describable as objects in space-time, but having from the outset your spinor qualities.

Thus I would consider quantum theory in the very narrow sense, as a theory which formally describes objects which are composites of the most simple objects: namely objects having just two dimensions in Hilbert space.

This way of looking at the matter, however, seems in a way to violate your statement that quantum theory is contained in a mixture of formal and informal language, because I leave out much of the informal language which already refers to space. Then, just as you and Aharanov said you could learn something additional for informal language from the mathematical formulation of the formal language. What we retain of the informal language of quantum theory in such a scheme would be very much what is contained in Drieschner's presentation,‡ which immediately leads to such a finite dimensional Hilbert space. It does not lead, to begin with, to ordinary space.

(BOHM): I think there is definitely some mathematical relationship between what Drieschner talked about and this. However, we

† Part IV of this book.
‡ Drieschner spoke in the colloquium, but his material is presented in von Weizsäcker's paper.

differ in our use of language about the description of the situation. The use of the word 'object' in this context is not really appropriate. These are different descriptions, and although Drieschner has referred to objects, it is really a kind of description. The difficulty comes from the connotation of the word 'object' in the informal language. It is better to rule it out, and it would help guide the informal side of our thinking better.

(VON WEIZSÄCKER): Would you accept 'system' instead?

(BOHM): No, the same objection applies to 'system'.

(BASTIN): Before you have sets of objects I agree you haven't got an object at all. I was going to ask how you introduce a concept which might be identified with more than one object.

(BOHM): What we usually call objects are content introduced into the description just as you introduce content into the cartesian frame of analytical geometry. We have got to discuss how we bring content into this description. We are discussing the most primitive frame of description, and this is the hardest thing of all to change. We tend to be attached to our descriptive gestalts very firmly, and tend to rule out other gestalts as irrelevant. We attach the implicit meaning that this *gestalt* is what is.

Ordering is basically description; then this description is content, and this content is a higher order of ordering, and you start developing a hierarchy that way. This is a notion which I have barely begun to touch, but I would say that we are not very far on until we bring in the hierarchical ordering of ordering. There is not only ordering, but the ordering itself is subject to ordering—you see this all the time in structure. Structure *is* the ordering of ordering. Now we have a language of description, which is able to begin this, but if you try to say there are objects it gets in the way of hierarchical development.

I am also encouraged by the fact that chiral ordering has shown itself to be relevant in physics, whereas the Cartesian language would suggest that it should be irrelevant. This is a puzzle, and has been one of the most significant developments in physics since the Dirac equation—namely the discovery that chiral ordering—the difference between right and left hand—is relevant in physics.

(BASTIN): What do you mean by 'an ordering'? I know what it usually means but the way you use it is more general. Can you throw yourself outside your normal way of thinking sufficiently to see the puzzle it is to anybody else?

(BOHM): The puzzle arises possibly because we are bound to have primitive terms in the language and we cannot possibly explain every term in terms of other terms. I say ordering is a primitive term. If you take any concept, whether measure or length, etc., it presupposes ordering. Without ordering there is no measure; there is not anything that you can do or communicate about. Let's say that ordering is a primitive term which has no fixed content.

(BASTIN): But some things must be more like it than others: for example it must be more like the case of a child putting down an alphabet than like...

(BOHM): He is experimenting with ordering, but it's a very special case.

(BASTIN): But it is more the case that he is experimenting with ordering then than when he breaks panes of glass.

(BOHM): That is an experiment with ordering, it is another kind of ordering—called *dis*-ordering. This difference between ordering and disordering is also a cause of confusion. There is no such thing as disorder: whatever happens is ordering.

(LINNEY): Surely there are two things wrong with what you have said. If *everything* is ordering, the point you are trying to make about ordering is lost because there is only some point if there is some contrast with disorder. The other point, which is more serious, is your general remark (which seems to pass without comment) that because it is a primitive notion you can't say anything about it. You can, in fact, have a series of primitives all defined in terms of each other, and this is an important and valuable exercise because it clears up the relation of these things to each other and it may teach you a great deal.

(VON WEIZSÄCKER): You said there is no disorder? I accept that as a starting point. That would mean that you use the term 'order' to describe what is, and what can be understood, and you say there is nothing which would not be order. In this sense, your term order is co-extensive with existence (not identical but co-extensive) and you say what you can understand is order. This is precisely the starting point of Plato's philosophy. He called 'form' what you call 'order'. What you try to do here is give a general abstract order dissociated from the particular content you wish to put into it. This is certainly a possibility. It is what is behind the formulation of such a concept as space. The concept 'space', which is not at all a self-evident concept in its abstract sense, is really an invention of modern times. Greek philosophy and Greek mathematics didn't know the meaning of what we call 'space' and didn't need to. However you can invent such a thing as space, which then would be an abstract order, which is supposed to be there independently of what we put into it. And what you propose here, is just another order of the same generality, but which is not identical with what we have classically called 'space'.

(BOHM): Whatever is described is always an order, however complicated it may be.

(BASTIN): You do wish to call it order and not form however, and I think for good reason.

(VON WEIZSÄCKER): Why should order be better than form? Why do you think he prefers order to form?

(BASTIN): Basically I think because he is going to introduce temporality at a fairly early stage.

(LINNEY): I think there is a more important point. I think it is because, whatever he says, and whatever he does, at some point he is also going to get out of it all the kinds of order that people familiarly cover by the word order. If he weren't going to consider such things as symmetry, asymmetry, etc., he wouldn't use the notion of order at all. Also, I agree with Bastin that this motivation is connected with what I believe is a fundamental kind of order—temporal order.

DEFINABILITY AND MEASURABILITY
IN QUANTUM THEORY†

YAKIR AHARONOV AND AAGE PETERSEN

1. The mathematical formalism of quantum mechanics imposes restrictions on the definability of physical variables in a given state of a system. The relation of these restrictions to the possibilities of measuring the variables has been a main issue in the discussion of the foundations of quantum theory.

Heisenberg [1] and Bohr [2] claimed that there is a complete agreement between the possibilities of definition and the possibilities of measurement. Against this view, Einstein [3] repeatedly tried to show that measuring procedures are available which permit us to specify physical variables with greater accuracy than that allowed by the quantal formalism. In contrast, Landau and Peierls [4] argued that in quantum field theory the methods of measurement are far more restrictive than the limitations imposed by the formalism. However, in each case [5], [3] a closer analysis of the measurement problem restored the equivalence between definability and measurability.

In the discussion referred to above the measurement problem was discussed by considering specific gedankenexperiments. A systematic investigation of the relation between definability and measurability must be based on the canonical description of measurements. In the quantal formalism the state of a system is represented by a wave function ψ or by a complete set of commuting hermitian operators of which ψ is an eigenfunction. To establish the equivalence between the mathematical description and the measurement possibilities one would have to show that any complete set of commuting hermitian operators, and no more, can be measured jointly with arbitrary accuracy. The canonical description of such measurements was given by von Neumann. [6] The system and apparatus is described by a Hamiltonian that includes a number of interaction terms each of which is proportional to one of the hermitian operators of the commuting set as well as to a suitable hermitian operator of the apparatus. These interactions are turned on for a short period of time, and when they are over, a correlation has been established between the apparatus and the system such that by 'reading' the apparatus one may infer the state of the system.

† This work was supported by Grant No. AFCRL-F-19 628-68-C-0053.

2. Despite the strong support that the Heisenberg–Bohr principle of equivalence between definability and measurability received in the early phase of the discussion of the foundation of quantum mechanics, this principle has remained a controversial issue. Thus, Fock and others[7] have argued that the von Neumann analysis is not satisfactory and that in addition to the limitations on measurements exhibited by the canonical formalism there exist restrictions which may be found by direct analysis of gedankenexperiments. An example of such additional restrictions is provided by the gedanken experiments illustrating the uncertainty relation between time and energy.

As is well known, the time–energy uncertainty relation cannot be deduced from the commutation relations in the usual way, since the time is not a dynamical variable but a parameter. This has given rise to two different interpretations of the meaning of Δt. According to the first, Δt refers to the uncertainty in any dynamical 'time' defined by the system itself; for example, the position of the hand of a clock is such a dynamical variable. If the energy of the clock has been measured with an accuracy ΔE, then there must be an uncertainty in the position of the hand such that the corresponding $\Delta t \geqslant h/\Delta E$. According to the second interpretation, Δt refers to the period during which the energy measurement takes place. In other words, the uncertain time is not related to any dynamical variable belonging to the system itself but rather to the laboratory time which specifies when the energy is measured.

While the first interpretation of the time–energy uncertainty relation is in complete agreement with the mathematical description of the system, the second interpretation is not. To see this, note that we may write down a quantum state corresponding to a well-defined energy at a given time; therefore the energy can be defined at a given laboratory time. Thus, if the second interpretation were correct, we would have an example of a discrepancy between definability and observability in non-relativistic quantum theory.

The second interpretation was suggested by the gedanken experiment that was originally used to illustrate time–energy measurements, i.e. a collision between the object and the test body. However, already Bohr and Rosenfeld[5] discovered that this gedanken experiment does not reflect the optimal measurement possibilities and that there exists another experiment, involving two collisions, in which the energy is measured with arbitrary accuracy at an arbitrarily well-defined laboratory time. The same result was found independently[8] by using the canonical description of measurements. On this method there corresponds to each formal interaction a gedanken experiment, and the interaction appropriate for an energy measurement immediately showed the necessity of a double instead of a single collision.

However, even if the canonical approach to measurements is considered satisfactory, the question may still be raised whether the equivalence between definability and measurability has been demonstrated conclusively. If there are no restrictions whatsoever on the interactions describing measurements, then the answer is positive, since in that case every hermitian operator would be measurable. But it is well known that even in non-relativistic theory there are restrictions on interactions imposed by symmetry principles or associated conservation laws. For example, a permissible interaction must be invariant under translations, rotations, gauge transformations, etc. To each of these restrictions on interactions there corresponds a restriction on measurements. For example, translation invariance implies that no interaction exists which depends on the position conjugate to the momentum that generates the translation. For this position to be measurable, an interaction dependent on the position must exist. Thus, the position is not measurable. Similarly, gauge invariance, which is equivalent to charge conservation, implies that the phase conjugate to charge cannot be measured, etc.

The limitations on interactions imposed by conservation laws have been construed as a violation of the principle of equivalence between definability and measurability in quantum mechanics.[9] However, such an interpretation is misleading,[10] for the same conservation laws force us to re-examine the meaning of the physical variables that appear in the formalism. In view of the invariance principles these variables cannot be given any absolute meaning but must be interpreted as *relative* to some frame of reference. Such relative variables can be measured with any accuracy without violating the invariance principles. For example, in measuring the position of a particle relative to a reference frame provided by a microscope, the total momentum of particle and microscope is conserved. The relevant interaction is invariant under a translation involving both particle and microscope and is, therefore, allowed. Thus, the restrictions stemming from invariance principles have no special bearings on measurements in non-relativistic quantum mechanics, and in particular they do not destroy the equivalence between definability and measurability in that domain.

3. While the Heisenberg–Bohr principle seems to be valid in non-relativistic quantum mechanics, it does not seem to be valid in the current formulation of relativistic quantum theory. As is well known, the causality condition imposes a new severe restriction on interactions: only local interactions exist.

In classical relativistic theory, this restriction on interactions does not impose limitations on measurements. Local interactions correspond to coincidence experiments. Such experiments are obviously sufficient to measure the position of a particle. Further, the velocity of a classical particle may be found by two successive position measurements. Thus, in classical theory, local interactions are sufficient to measure the state of a particle, and indeed of any classical system. We see that the relativistic constraint on interactions causes no limitations on measurements in the classical domain because classically we can measure the velocity of a particle by using two kinds of interactions. One of these depends on the velocity; the other depends on the position and is switched on twice. The relativistic constraint excludes the first, but leaves the second available.

In quantum theory the situation is different. Because of the lack of commutativity of position and momentum, we cannot use the second kind of interaction to measure velocity (at least not in short periods of time). Thus, it is not surprising that the relativistic constraint on interactions imposes limitations on measurements in the quantum domain. For example, to measure the momentum of a field according to the canonical method one would need an interaction proportional to the momentum and to some external degree of freedom belonging to the apparatus. It is easy to show that such an interaction is not causal and therefore not permissible. To preserve causality, one is forced to measure momentum in an indirect way, e.g. by measuring the position twice, and any such indirect procedure implies that if the intended accuracy is Δp, then the measurement must last at least a period T, where $T \geqslant h/c\Delta p$. In fact, this is but one example of a limitation on measurements due to the combination of the restrictions imposed by the quantum and the velocity of light.

This new type of limitation, pertaining to measurement of a single variable rather than two conjugate variables, has no counterpart in the current mathematical formulation of relativistic quantum theory. For example, at present there are no formal reasons why one cannot write down a state of a field at a given time that has a well-defined total momentum. Such a state, however, involves more information than can be obtained by measurement. Thus, in relativistic quantum theory there is a discrepancy between definability and measurability.

It may be noted that the discrepancy is removed if one considers only asymptotic states, since in that case the time available for measurement is not restricted and the relativistic constraint loses its significance. Thus, if one considers the discrepancy an indication that the current relativistic quantum theory should be modified, it seems that this modification would not affect current S-matrix theory but only short-time aspects of the theory.

REFERENCES

(1) Heisenberg, W. *The Physical Principles of the Quantum Theory* (Chicago, 1930). See also Heisenberg's article in S. Rosental (ed.), *Niels Bohr* (Amsterdam, 1967), 92.
(2) Bohr, N. *Atomic Theory and the Description of Nature* (Cambridge, 1961).
(3) See the description of the discussions between Einstein and Bohr in P. A. Schilpp (ed.), *Albert Einstein, Philosopher-Scientist* (Evanston, 1949), 199.
(4) Landau, L. and Peierls, R. *Z. Phys.* (1931) **69**, 56.
(5) Bohr, N. and Rosenfeld, L. *Dan. Vid. Selsk. Math.-fys. Medd.* (1933) **12**, 8.
(6) von Neumann, J. *Mathematical Foundations of Quantum Mechanics* (Princeton, 1955), chap. VI.
(7) See, for example, V. Fock and N. Krylov, *J. Phys.* (U.S.S.R.) (1947) **11**, 112.
(8) Aharonov, Y. and Bohm, D. *Phys. Rev.* (1961) **122**, 1649.
(9) See, for example, E. P. Wigner, *Z. Phys.* (1952) **131**, 101; H. Araki and M. M. Yanase, *Phys. Rev.* (1960) **120**, 622; and G. C. Wick, A. S. Wightman and E. P. Wigner, *Phys. Rev.* (1952) **88**, 101.
(10) See also A. Aharonov and L. Susskind, *Phys. Rev.* (1967) **155**, 1428.

THE BOOTSTRAP IDEA AND THE FOUNDATIONS OF QUANTUM THEORY†

GEOFFREY F. CHEW

Thirty years of intensive effort have failed to produce a theoretical structure reconciling the axioms of quantum mechanics with principles of special relativity. No demonstration has been given of absolute irreconcilability, but the clouded situation adds to the doubts surrounding the foundations of quantum mechanics. Recent experimental and theoretical developments in hadron research [1] suggest that the origin of the hadrons may lie in an intrinsically relativistic mechanism called the 'bootstrap'. In this note I review the bootstrap idea and argue its incompatibility with a certain basic feature of quantum theory.

An oft heard version of the bootstrap hypothesis is that all hadrons are 'composites' of each other, none being elementary. Each hadron plays three different roles: it may be a 'constituent' of a 'composite structure', it may be 'exchanged' between constituents and thus constitute part of the force holding the structure together, or it may itself *be* the entire composite. A familiar picture qualitatively describes a meson as a baryon-antibaryon composite, held together by exchange of mesons, while at the same time a baryon is a meson-baryon composite held together by exchange of baryons. Pictures of this type are often used but are unacceptably vague, suffering from dependence on conventional nonrelativistic dynamical language in a situation in which relativity is crucial. The bootstrap mechanism is unavoidably relativistic because the binding energy in the composite must be comparable to the rest-mass energies of constituent particles.

As stated above, there exists at present no mechanical framework consistent with both quantum and relativistic principles. The chief candidate is local Lagrangian field theory, but countless theoretical studies have suggested insuperable pathology in the concept of interaction between fields at a point of space-time. Furthermore, although tentative bootstrap investigations often have employed Lagrangian models, we shall see below that this framework is intrinsically un-

† This work was supported in part by the U.S. Atomic Energy Commission.

suited to the bootstrap idea. An alternative is the analytic S-matrix,[2] abandoning not only conventional Dirac quantum mechanics but even a meaning for microscopic space-time in describing interactions between hadrons.

The basic S-matrix concept is momentum, measured for freely moving hadrons before and after their collisions with each other. No effort is made to describe the collision itself. Each element of the S-matrix describes a particular nuclear reaction and depends on the momenta and spins of the initial and final hadrons participating in the reaction. The experimental definition of momentum and spin involves macroscopic space-time, but no precise meaning need be given to the 'position' of a hadron.

In S-matrix theory there is no 'state-vector' that evolves in time and no operators associated with observables. Nevertheless, since elements of the S-matrix encompass all conceivable hadron experiments, ability to predict this matrix would constitute a complete hadronic theory. The essence of the bootstrap conjecture is that three or four general constraints, each with persuasive experimental motivation, suffice to define a unique S-matrix. The first constraint is on macroscopic space-time, combining the familiar Lorentz (or Poincaré) invariance with the 'cluster' requirement that reactions well separated in space-time are independent. The second is unitarity, which combines superposability of free-particle amplitudes with conservation of probability. The third constraint is more subtle and is related to the nonexistence of zero-mass hadrons. Experiments suggest that S-matrix elements are *analytic* functions of the hadron momenta on which they depend. In the absence of zero-mass hadrons there is no bar to complex continuation, apart from isolated singularities required by unitarity. Among these singularities, simple poles correspond to particles, the pole position determining the particle mass while the residue determines a partial width. (A complex mass means that the hadron is unstable, the imaginary part of the mass corresponding to the lifetime.) Also occurring is a variety of branch points associated with the name of L. D. Landau (for a reasonably up-to-date survey see Eden, Landshoff, Olive and Polkinghorne[2]) and related to the possibility of complicated reactions proceeding through a succession of simpler reactions. Causality is ensured by the proper location of the Landau branch points.[3] The S-matrix constraint of 'first-degree analyticity' postulates no momentum singularities other than particle poles and Landau branch points. This third constraint possesses substantial experimental support, although its basis is not as compelling as that for the constraints of Lorentz invariance and unitarity.

The pole-particle correspondence fails to distinguish between 'elementary' and 'composite' particles, but ten years ago in non-relativistic potential-scattering theory Tullio Regge identified a connection between composite-particle poles and S-matrix behaviour as certain momenta approach infinity. Unitarity demands some such connection for relativistic high-spin hadrons, and it was conjectured by S. Frautschi and the author that Regge asymptotic behaviour might be used in the relativistic hadron S-matrix to *define* 'compositeness'. Thus a possible fourth S-matrix constraint is that all poles should be Regge poles, a condition sometimes called 'second-degree analyticity' because Regge behaviour involves analytic continuation in angular as well as in linear momenta. For more details see Chew.[4] A different term often applied to the same condition is 'nuclear democracy'. This fourth constraint had little experimental support when it was proposed in 1961 but it has by now acquired respectability. Although there exists no firm logical bar to alternative constraints involving elementary particles without arbitrary parameters, the aesthetic principle of 'lack of sufficient reason' may be invoked. There is no 'need' for elementary hadrons.

In summary, the bootstrap hypothesizes that observed hadron phenomena correspond to the *unique* Lorentz-invariant, unitary, analytic S-matrix containing only Regge poles. There is supposed to exist no fundamental 'building block,' field or particle.

A possible path for experimental demolition of the bootstrap hypothesis would be the discovery of non-Regge poles among hadrons. Established universality of Regge poles among hadrons, however, would not convince all physicists that we are dealing with a bootstrap. It is often conjectured that beneath nuclear matter lies a basic field, unconnected with special hadrons and obeying a simple 'master equation' of motion. Lagrangian models indicate a tendency for hadrons sharing the quantum numbers of such a master field to exhibit 'aristocratic' properties, such as not lying on a Regge trajectory, but no general demonstration ever has been given that a master equation precludes nuclear democracy. Since a master field would be inaccessible to direct measurement, one may despair of ever verifying its actuality.

Although it seems difficult to formulate an experimental test to distinguish between the bootstrap and such a field theory, a general theoretical distinction is possible even without specification of the master equation. Field theory rests on the 'closure' aspect of Dirac quantum mechanics, that is, on the existence of an 'inner Hilbert space' of states with respect to which the product of two operators may be defined. In S-matrix theory, on the other hand, there are no Dirac operators and correspondingly no immediate need for closure. If a master field underlies the S-matrix, closure must have some meaning, but we propose now to argue that the bootstrap idea precludes any significance.

To avoid confusion, it must be recognized that S-matrix unitarity implies closure within subsets of free-particle states sharing a common value of energy-momentum. These asymptotic states in a formal sense can be said to constitute an 'outer' Hilbert space but, except in a nonrelativistic approximation that preserves the number of each type of particle, there is no reason to expect physical equivalence between 'inner' and 'outer' spaces.

Our basic argument is that conventional quantum-mechanical closure of the 'inner' Hilbert space implies a distinguishable set of degrees of freedom for the dynamical system. We here use the term 'degrees of freedom' to describe a 'minimum' set of operators through repeated applications of which the entire inner Hilbert space may be spanned. The degrees of freedom may be infinite and may be continuous, but closure requires a sense in which a minimal set can be identified. A fundamental field provides such a sense; the fact that the manifestations of such a field may include an infinite number of composite particles should not be confused with the identifiability of its degrees of freedom, a characteristic of any system described by conventional quantum mechanics. Conversely, such identifiability is physically inseparable from the general concept of closure.

Modern approaches to field theory avoid mentioning the 'degrees of freedom' concept, but without it the notion of closure (or completeness) of the inner Hilbert space is unnatural. That is to say, the 'size' of the inner Hilbert space may be associated with the number of underlying degrees of freedom. If the number of degrees of freedom is increased, the 'size' of the space increases and conversely. In this way of thinking, the notion of 'spanning the space' is tied to the 'size' of the space and thus to the number of degrees of freedom.

Now an essential aspect of the bootstrap idea is to abandon meaning for a definite underlying set of dynamical degrees of freedom. Were such to exist, they would constitute 'building blocks'. This point tends to be overlooked because of the evident freedom, bootstrap or no, enjoyed by the experimenter in setting up initial configurations of stable particles. The relativistic experimenters, however, lack the capacity to set up a 'quantum state' in the sense of Dirac, where a meaning is attached to subsequent state-evolution from one instant of time to another. The capabilities of physical measurement require meaning only for free-particle configurations, and it begs the question to assume that the flexibility available to a nuclear experimenter corresponds to the degrees of freedom of an inner Hilbert space.

Because the 'degrees of freedom' notion never rears its head within the S-matrix framework, this structure is suitable to formulation of the bootstrap. Confusion, indeed, arises when unitarity sums are truncated in approximate S-matrix calculations; certain particles are thereby in effect given an 'elementary' status and provide a basis for 'counting'. This phenomenon associated with approximation has important utility in special situations where unusual mass ratios occur. A prime example is the small pion mass, compared with that of neutrons and protons, which allows an approximate meaning for nucleon wave functions in nonrelativistic nuclear physics. In such special situations one may speak of 'approximate closure', the goodness of the approximation being governed by the mass ratios and the smallness of velocities; but no basis is provided by S-matrix theory for any general notion of closure, which requires an identifiable set of degrees of freedom. The essence of the bootstrap notion in foregoing any master equation is that there *are* no degrees of freedom! More accurately, the 'degrees of freedom' concept is inadmissible.

Lacking the 'degrees of freedom' concept, the concept of a state vector for the system would make no sense. In simple nonrelativistic situations a certain *approximate* sense could be achieved, but the approximation might not be sufficiently accurate to justify extending the idea of a 'quantum state' to large and complicated systems.

REFERENCES

(1) Chew, G. F., Gell-Mann, M. and Rosenfeld, A. *Scient. Am.* (February 1964).
(2) Eden, R. J., Landshoff, P. V., Olive, D. I. and Polkinghorne, P. C. *The Analytic S-Matrix* (Cambridge University Press, 1966).
(3) Chandler, C. and Stapp, H. P. *Macroscopic Casuality Conditions and Properties of Scattering Amplitudes* (Lawrence Radiation Laboratory Report UCRL-17734, 14 June 1967, submitted to *J. math. Phys.*).
(4) Chew, G. F. *The Analytic S-Matrix* (New York: W. A. Benjamin and Co., 1966).

V

A FRESH START?

The average physicist imbibes from the very start of his training a feeling for the immutability of his science. He knows that if you change the fundamentals in one particular you have got to change everything, and since the scheme as we have it is both massively coherent and massively successful, he tends to conclude that you can't change anything. In the realm of the atomic, this feeling is particularly strong and in Cambridge we used to call it 'Cavendish Commonsense'. The more philosophical tradition, by contrast, was always very sensitive to the problem of settling empirically the rival claims of different formal scientific systems, and tended to presuppose that it was commonplace to have several lying around to hand. They, the philosophers, didn't know that when you had reached that picking and choosing situation, you had, from the physicists' point of view, a problem that had become technical merely, if not actually trivial.

The clash of the traditions can be illustrated by the case of electron spin. For the physicist the electron is just a spinning body like any other except in certain details. For the philosopher, however, electron spin was a theoretical concept having merely a metaphorical relation with cricket ball spin, and he turned an incredulously blind eye to the fact that if you select electrons of one spin-sign and fire them into a lump of lead, that lump of lead will start to rotate.

These have been the traditions. Unfortunately they diverged, each into its own compartmentalized rigidity. However, for this colloquium it seemed that there was sufficient conceptual confusion in the foundations of quantum theory to make it worthwhile searching for people prepared to put forward radical alternatives in order to assess just how difficult such a task really is. Also one wants to know just how obscure is the void in which Bohr—sophisticatedly—and Cavendish Commonsense—crudely—assure us we shall find ourselves if we abandon the classical physical concepts.

The stimulus to set up basic new forms, as far as participants in the colloquium were concerned, was predominantly the unsatisfactory relation of the discrete character of atomic systems to the

147

classical continuum. Preoccupation with discreteness first made itself felt in Chew's paper in the foregoing part of the book. Penrose continues the theme, in this, with discrete angular momentum in which points can be constructed as required. Preoccupation with discreteness will also reappear in our last, more philosophical, Part VI, where von Weizsäcker and Drieschner investigate the consequences of a finitistic modification to the logic of quantum states and Hilbert space.

Penrose's paper led to a good deal of discussion. Topics raised were: his approach to the continuum, his assumption of rules derived from current quantum theory, and his use of three-dimensional space. The paper which actually appears in Part V has been clarified with these points in view. Nevertheless since in it ideas extracted from existing quantum theory are combined with a fundamentally discrete method, it is likely that the attention of critics will continue in the main to be directed to the points at which these seemingly disparate sets of ideas have to make contact.

Participants would naturally have liked a simple answer to the question 'did Penrose put the space-time structure into his theory or did he deduce it from the theory?' Certainly he took the commutation rules of quantum theory as his first inspiration in choosing rules for associating angular momentum, and there was no agreed point of view on the extent to which these presuppose the space-time structure. (It is a fascinating question, though.)

The continuity perplexity was this: was Penrose using the continuum to define the continuum or was he not? Von Weizsäcker remarked that Penrose was building space-time by making use of probabilities which themselves are described by real numbers and which, in Penrose's case, just happen to be rational. Penrose's explicit method, by contrast, was to think of probabilities as arising in 'some more primitive theory from choices between finite alternatives', and to generate as many points in angular momentum space as are operationally required for any specified purpose. However, at some point there has to be an assimilation of Penrose's probabilities to those current in quantum theory, and at the moment probabilities defined in the field of real numbers are what are used in quantum theory, and their use is integral to the current operational interpretation of that theory.

Later papers in this part of the book are more radical in their discrete approach than Penrose's, and their authors paid the penalty in presenting a great barrier to discussion.

Atkin sets up a model of the observation process in which the 'scale' or comparator-system is always a finite ordered set and in which progressive refinement of observation must be accompanied by the inclusion of more points in the scale. Hiley introduces this topic with a survey of possible approaches to a discrete phase space—especially in regard to the quantum theoretical limitation on simultaneous momentum and position measurement.

The experimental identification of the elements of a simplicial complex caused a good deal of discussion; as these did not lead to an unambiguous conclusion some attempts were made to discover analogues in more familiar theory for some of the theoretical constructions of Atkin and Hiley to see if these could be an alternative starting point. It seemed likely—for example—that commutation relations might play this part.

Bastin shows that a concept of discrimination between physical entities, used as the basic logical operation, can itself generate the larger scale sets of the type required by Atkin, and that the resulting algebraic model throws light on the origins of the concept of spin in atomic structure. His paper provoked a suggestion from Aharonov that computing methods of the sort he was discussing might be used to bridge the gap—which Part III on measurement theory had shown to be very serious—between the individual quantum system and the quantum assembly.

ANGULAR MOMENTUM: AN APPROACH TO COMBINATORIAL SPACE-TIME

ROGER PENROSE

I want to describe an idea which is related to other things that were suggested in the colloquium, though my approach will be quite different. The basic theme of these suggestions has been to try to get rid of the continuum and build up physical theory from discreteness.

The most obvious place in which the continuum comes into physics is in the structure of space-time. But, apparently independently of this, there is also another place in which the continuum is built into present physical theory. This is in quantum theory, where there is the superposition law: if you have two states, you're supposed to be able to form any linear combination of these two states. These are complex linear combinations, so again you have a continuum coming in—namely the two-dimensional complex continuum—in a fundamental way.

My basic idea is to try and build up both space-time and quantum mechanics simultaneously—from *combinatorial* principles—but not (at least in the first instance) to try and change physical theory. In the first place it is a *reformulation*, though ultimately, perhaps, there will be some changes. Different things will suggest themselves in a reformulated theory, than in the original formulation. One scarcely wants to take *every* concept in existing theory and try to make it combinatorial: there are too many things which look continuous in existing theory. And to try to eliminate the continuum by *approximating* it by some discrete structure would be to change the theory. The idea, instead, is to concentrate only on things which, in fact, *are* discrete in existing theory and try and use them as *primary concepts*— then to build up other things using these discrete primary concepts as the basic building blocks. Continuous concepts could emerge in a limit, when we take more and more complicated systems.

The most obvious physical concept that one has to start with, where quantum mechanics says something is discrete, and which is connected with the structure of space-time in a very intimate way, is in *angular momentum*. The idea here, then, is to start with the concept of angular momentum—where one has a *discrete* spectrum—and use the rules for combining angular momenta together and see if in some sense one can construct the concept of *space* from this.

151

One of the basic ideas here springs from something which always used to worry me. Suppose you have an electron or some other spin $\frac{1}{2}\hbar$ particle. You ask it about its spin: is it spinning up or down? But how does it know which way is 'up' and which way is 'down'? And you can equally well ask the question whether it spins right or left. But whatever question you ask it about directions, the electron has only just *two* directions to choose from. Whether the alternatives are 'up' and 'down', or 'right' and 'left', depends on how things are connected with the macroscopic world.

Also you could consider a particle which has *zero* angular momentum. Quantum mechanics tells us that such a particle has to be spherically symmetrical. Therefore there isn't really any choice of direction that the particle can make (in its own rest-frame). In effect there *is* only one 'direction'. So that a thing with zero angular momentum has just one 'direction' to choose from and with spin one-half it would have *two* 'directions' to choose from. Similarly, with spin one, there would always be just three 'directions' to choose from, etc. Generally, there would be $2s+1$ 'directions' available to a spin s object.

Of course I don't mean to imply that these are just directions in space in the ordinary sense. I just mean that these are the choices available to the object as regards its state of spin. That is, however we may choose to interpret the different possibilities when viewed on a macroscopic scale, the object itself is 'aware' only that these are the different possibilities that are open to it. Thus, if the object is in an s-state, there is but one possibility open to it. If it is in a p-state there are three possibilities, etc., etc. I don't mean that these possibilities are things that from a macroscopic point of view we would necessarily think of as directions in all cases. The s-state is an example of a case where we would not!

So we oughtn't at the outset to have the concept of macroscopic space-direction built into the theory. Instead, we ought to work with just these discrete alternatives open to particles or to simple systems. Since we don't want to think of these alternatives as referring to pre-existing directions of a background space—that would be to beg the question—we must deal only with *total* angular momentum (j-value) rather than spin in a direction (m-value).

Thus, the primary concept here has to be the concept of *total angular momentum* not the concept of angular momentum in, say, the z-direction, because: which is the z-direction?

Imagine, then, a universe built up of things like that shown in fig. 1. These lines may be thought of as the world-lines of particles. We can view *time* as going in one direction, say, from the bottom of the diagram to the top. But it turns out, really, that it's irrelevant which way time is going. So I don't want to worry too much about this.

I'm going to put a number on each line. This number, the *spin-*

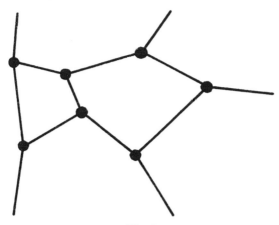

Fig. 1

number, will have to be an integer. It will represent twice the angular momentum, in units of $\hbar$. All the information I'm allowed to know about this picture will be just this diagram (fig. 2): the network of connections and spin numbers 3, 2, 3, ... like that. I should say

Fig. 2

that the picture I want to give here is just a model. Although it does describe a type of *idealized* situation exactly according to quantum theory, I certainly don't want to suggest that the universe 'is' this picture or anything like that. But it is not unlikely that *some* essential features of the model that I am describing could still have relevance in a more complete theory applicable to more realistic situations.

I have referred to these line segments as representing, in some way, the world-lines of particles. But I don't want to imply that these lines stand just for *elementary* particles (say). Each line could represent some compound system which separates itself from other such systems for long enough that (in some sense) it can be regarded as isolated and stationary, with a well-defined total angular momentum $n \times \frac{1}{2}\hbar$. Let us call such a system or particle an n-unit. (We allow $n = 0, 1, 2, \ldots.$) For the precise model I am describing, we must also imagine that the particles or systems are not moving relative to one another. They just transfer angular momentum around, regrouping themselves into different subsystems, perhaps annihilating one another, perhaps producing new units. In the diagram (fig. 2), the 3-unit at the bottom on the left splits into a 2-unit and another 3-unit. This second 3-unit combines with a 1-unit (produced in the break up of a 2-unit into two 1-units) to make a new 2-unit, etc., etc. It is only the *topological* relationship between the different segments, together with the spin-number values, which is to have significance. The time-ordering of events will actually play no role here (except conceptually). We could, for example, read the diagram as though time increased from the left to the right, rather than from the bottom to the top, say.

Angular momentum conservation will be involved when I finally give the rules for these diagrams. These rules, though combinatorial, are actually derived from the standard quantum mechanics for angular momentum. Thus, in particular, the conservation of total angular momentum must be built into the rules.

Now, I want to indicate answers to two questions. First, what *are* these combinatorial rules and how are we to interpret them? Secondly, how does this enable us to build up a concept of space out of total angular momentum? In order not to get bogged down at this stage with too much detail, I shall defer, until later on, the complete definition of the combinatorial rules that will be used. All I shall say at this stage is that every diagram, such as fig. 2 (called a *spin-network*) will be assigned a non-negative integer which I call its *norm*. In some vague way, we are to envisage that the norm of a diagram gives us a measure of the frequency of occurrence of that particular spin-network in the history of the universe. This is not actually quite right—I shall be more precise later—but it will serve to orient our thinking. We shall be able to use these norms to calculate the probabilities of various spin values occurring in certain simple 'experiments'. These probabilities will turn out always to be rational numbers, arising from the fact that the norm is always an integer. Given any spin-network, its norm can be calculated from it in a purely combinatorial way. I shall give the rule later.

But first let me say something about the answer to the second question. How can I say anything about *directions* in space, when I only have the non-directional concept of *total* angular momentum? How do I get 'm-values' out of j-values, in other words?

Clearly we can't do quite this. In order to know what the 'm-value' of an n-unit is, we would require knowledge of which direction in space is the 'z-direction'. But the 'z-direction' has no physical meaning. Instead, we may ask for the 'orientation' of one of our n-units *in relation to some larger structure* belonging to the system under consideration. We need some larger structure which in fact does give us something that we may regard as a well-defined 'direction in space' and which could serve in place of the 'z-direction'. As we have seen, a structure of spin zero, being spherically symmetrical, is no good for this; spin $\frac{1}{2}\hbar$ is not much better; spin $\hbar$ only a little better; and so on. Clearly we need a system involving a fairly large total angular momentum number if we are to obtain a reasonably well-defined 'direction' against which to test the 'spin direction' of the smaller units. We may imagine that for a large total angular momentum number N, we have the potentiality, at least, to define a well-defined direction as the *spin axis* of the system. Thus, if we *define* a 'direction' in space as something associated with an N-unit with a large N value (I call this a *large unit*), then we can ask how to define *angles* between these 'directions'. And if we can decide on a good way of measuring angles, we can then ask the question whether the angles we get are consistent with an interpretation in terms of directions in a Euclidean three-dimensional space, or perhaps in some other kind of space.

How, then, are we to define an angle between two large units? Well, we do this by performing an 'experiment'. Suppose we detach a 1-unit (e.g. an electron, or any other spin $\frac{1}{2}\hbar$ particle) from a large N-unit in such a way as to leave it as an $(N-1)$-unit. We can then re-attach the 1-unit to some other large unit, say an M-unit. What does the M-unit do? Well (according to the rules we are allowed here) it can either become an $(M-1)$-unit or an $(M+1)$-unit. There will be a certain probability of one outcome and a certain probability of the other. Knowing these probability values, we shall have information as to the angle between the N-unit and the M-unit. Thus, if our two units are to be 'parallel', we would expect zero probability for the $M-1$ value and certainty for the $M+1$ value. If the two units are to be 'antiparallel' we would expect exactly the reverse probabilities. If they are 'perpendicular', then we would expect equal probability values of $\frac{1}{2}$, for each of the two outcomes. Generally, for an angle θ between the directions of the two large units we would expect a

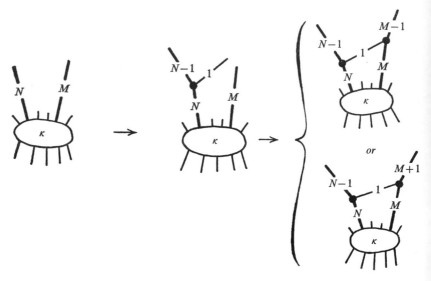

Fig. 3

probability $\frac{1}{2} - \frac{1}{2}\cos\theta$ for the M-unit to be reduced to an $(M-1)$-unit and a probability $\frac{1}{2} + \frac{1}{2}\cos\theta$ for it to be increased to an $(M+1)$-unit.

Let me draw a diagram to represent this experiment (fig. 3). Here κ represents some known spin-network. By means of a precise (combinatorial) calculational procedure—which I shall describe shortly—we can calculate, from knowledge of the spin-network κ, the probability of each of the two possible final outcomes. Hence, we have a way of getting hold of the concept of Euclidean angle, starting from a purely combinatorial scheme.

As I remarked earlier, these probabilities will always turn out to be *rational* numbers. You might think, then, that I could only obtain angles with rational cosines in this way. But this would be a somewhat misleading way of viewing the situation. With a finite spin-network with finite spin-numbers, the angle can never be quite well-enough defined. I *can* work out numerical values for these 'cosines of angles' for a finite spin-network, but these 'angles' would normally not quite agree with the actual angles of Euclidean space until I go to the limit.

The view that I am expressing here is that rational probabilities are to be regarded as something which can be more primitive than ordinary real number probabilities. I don't need to call upon the full continuum of probability values in order to proceed with the theory. A rational probability $p = m/n$ might be thought of as arising because the universe has to make a choice between m alternative possibilities of one kind and n alternative possibilities of another—all of which are to be equally probable. Only in the limit, when numbers go to infinity do we expect to get the full continuum of probability values.

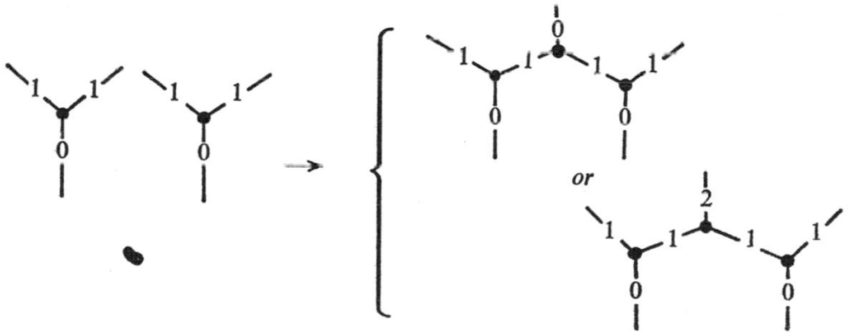

Fig. 4

As a matter of fact, it was this question of rational values for primitive probabilities arising in nature, which really started me off on this entire line of thought concerning spin-networks, etc. The idea was to find some situation in nature which one might reasonably regard as giving rise to a 'pure probability'. I am not really sure whether it is fair to assume that 'pure probabilities' exist in nature, but by these I mean probabilities (necessarily quantum mechanical) whose values are determined by nature alone and not in principle influenced by our ignorance of initial conditions, etc. I suppose I might have thought of branching ratios in particle decays as a possible example. Instead, I was led to consider a situation of the following type.

Two spin zero particles each decay into pairs of spin $\frac{1}{2}\hbar$ particles. Two of the spin $\frac{1}{2}\hbar$ particles then come together, one from each pair, and combine to form a new particle (fig. 4). What is the spin of this new particle? Well, it must be either zero or $\hbar$, with respective probabilities $\frac{1}{4}$ and $\frac{3}{4}$ (assuming no orbital components contribute, etc.). Although you can see that there are objections even here to regarding this as giving a 'pure probability', at least the example served as a starting point. (This example was to some extent stimulated by Bohm's version of the Einstein–Rosen–Podolsky thought experiment, which it somewhat resembles.) The idea, then, is that any 'pure probability' (if such exists) ought to be something arising ultimately out of a choice between equally probable alternatives. All 'pure probabilities' ought therefore, to be rational numbers.

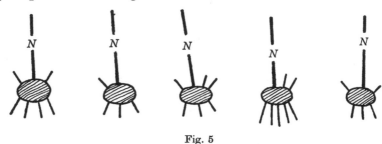

Fig. 5

But let me leave all this aside since it doesn't affect the rest of the discussion. Actually, I haven't quite finished my 'angle measuring experiment', so let me return to this.

Let us consider the following particular situation. Suppose we have a number of disconnected systems, each producing a large N-unit. There are to be absolutely no connections between them (fig. 5).

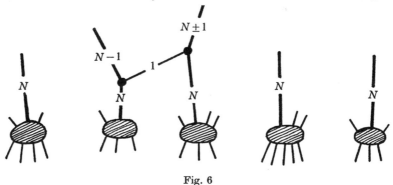

Fig. 6

Let me try to measure the 'angle' between two of them by doing one of the 'experiments' I described earlier. I detach a 1-unit from one of the N-units in such a way as to leave it as an $(N-1)$-unit. Then I reattach the 1-unit to one of the other N-units (fig. 6). According to the rules (cf. later) it will follow that the probability of the second

N-unit to become an $(N \pm 1)$-unit is $\frac{1}{2}(N+1\pm 1)/(N+1)$. These two probabilities become equal in the limit $N \to \infty$. Thus, if we are to assign an 'angle' between these units, then, for N large, this would have to be a *right-angle*. This is just using the probability blindly. I would similarly have to say, of any other pair of the N-units, that they are at right-angles. It would seem that I could put any number of N-units at right-angles to each other. In this instance I have drawn five. Does this mean that we get a five-dimensional space?—or an ∞-dimensional space?

Clearly I have not done things quite right. There are no connections between any of the N-units here, so one would like to think of the probabilities that arise out of one of these experiments as being not *just* due to the *angle* between the N-units (if they have an angle in some sense), but also due to the 'ignorance' implicit in the set-up. That is, we think of the probabilities as arising in two different ways. In the first instance, probabilities can arise in this type of experiment, if we have a *definite angle* between two spinning bodies (as we have seen). These are the genuine quantum mechanical probabilities. But, in the second instance, we may just be *uncertain* as to what the angle *is* between the two bodies. This lack of knowledge, concerning the history (or origins) of the two bodies, will give us a contribution to the probability value—an *ignorance* factor—which will serve to obscure the meaning of the probability in terms of angles. In the present instance, we are allowed absolutely no information concerning the interconnections between the different N-units, so the probability is

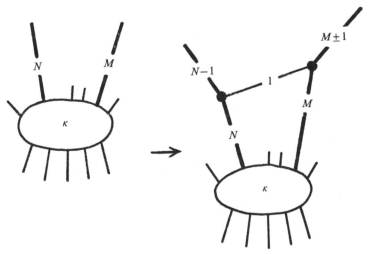

Fig. 7

not really due to 'angle' at all. In this extreme case, the probability is *entirely* 'ignorance factor'. In general, the two effects will be mixed up, so we shall need a means of separating them.

Let me change the picture a bit. I'll put in some 'known' connecting network (now denoted by κ) and have two large units coming out, as in fig. 7. I do one of these experiments, but then *repeat* the experiment. Suppose the N-unit is reduced to an $(N-1)$-unit and then to an $(N-2)$-unit. The M-unit becomes an $(M \pm 1)$-unit and then an $(M \pm 2)$-unit, or an M-unit again (fig. 8). The question is:

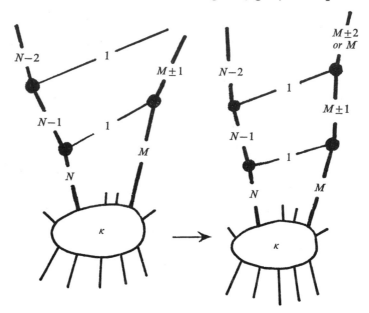

Fig. 8

is the probability of the second experiment influenced by the result of the first experiment? If this is essentially an 'ignorance' situation, where one doesn't initially know much about how the spin axes are pointing, then the *result* of the first experiment provides us with some *information* as to the relative directions of the spin axes. Therefore, the probabilities in the second experiment will be altered by the knowledge of the result of the first experiment.†

† It should be borne in mind that all these probability values are simply *calculated* here, from knowledge of the spin-networks involved. The 'experiments' are really theoretical constructions. However, it would be possible (in principle—with the usual reservations) to measure these probabilities experimentally, by simply repeating the experiment many times, each time reconstructing the spin-network afresh

If the probabilities calculated for the second experiment are *not* substantially altered by the knowledge of the result of the first experiment, then I say that the angle between the two large units is essentially well-defined. If they *are* substantially altered by the result of the first experiment, then there is a large 'ignorance' factor involved, and the probabilities arise not just from 'angle'.

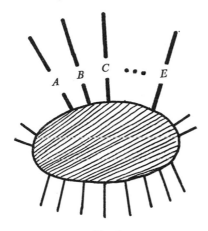

Fig. 9

Suppose, now, we have the system shown in fig. 9, which has a number of large units emerging, and suppose that it happens to be the case that the angle between any two of them is well-defined in the sense I just described. (All the numbers A, B, ... are large compared with unity; I can do a few odd experiments which do not much change these numbers.) Then there is a theorem which can be proved to the effect that these angles are all consistent with angles between directions in Euclidean three-dimensional space.

Now, should I be in any way surprised by this result? Admittedly I should have been surprised if the method gave me any different space; but on the other hand, it is not completely clear to me that the result is something I could genuinely have inferred beforehand. Let me mention a number of curious features of the theory in this context. In the first place, suppose I set the situation up with wave functions and everything, and work according to ordinary quantum mechanical rules. I have these particles (or systems) with large angular momentum, and I finally find out that I get these angles consistent with directions in Euclidean three-dimensional space. I never, at any stage, specified that these large angular momentum systems should, in fact, correspond to bodies which do have well-defined directions (as rotation axes). There are states with large total angular momentum (e.g. $m = 0$ states) which point all over the place, not necessarily in any one direction.

I can start from some *given* Euclidean 3-space and use an ordinary Cartesian description in terms of x, y, z. I can use particles (or systems) with large total angular momentum, but which do *not* happen to give well-defined directions in the original space. Then I work out the 'angles' between them and find that these angles do not correspond to anything I can see as angles in the original description, but they are nevertheless consistent with the angles between directions in *some* abstract Euclidean three-dimensional space. I therefore take the view that the Euclidean three-dimensional space that I get out of all this, using probabilities, etc. is the *real* space, and that the original space, with its x, y, z's that I wrote down, is an irrelevant convenience, like co-ordinates in general relativity, where one writes down any co-ordinates which don't necessarily mean anything. The central idea is that *the system defines the geometry*. If you like, you can use the conventional description to fit the thing into the 'ordinary space-time' to begin with, but then the geometry you get out is not necessarily the one you put into it. So I don't know whether I should be surprised or not by the fact that I actually get the right geometry in the end.

There is a second aspect of this work that I think I regarded as slightly surprising at first. This is the fact that although no complex numbers are ever introduced into the scheme, we can still build up the full *three*-dimensional array of directions, rather than, say, a two-dimensional subset. To represent all possible directions as states of spin of a spin $\frac{1}{2}\hbar$ particle, we need to take *complex* linear combinations (in the conventional formalism). Here we only use rational numbers—and complex numbers cannot be approximated by rational numbers alone! Again, the answer seems to be that the space I end up with is not really the 'same' space as the (x, y, z)-space that I could start with—even though both are Euclidean 3-spaces.

One might ask whether corresponding rules might be invented which lead to other dimensional schemes. I don't in fact see *a priori* why one shouldn't be able to invent rules, similar to the ones I use, for spaces of other dimensionality. But I'm not quite sure how one would do this. Also, it's not obvious that the whole scheme for getting the space out in the end would still work. The rules I use are derived from the irreducible representations of SO(3). These have some rather unique features.

Now, from what I've said so far, you might wonder whether you could just scatter the numbers on the network at random. Actually, you can if you like, but unless you are a bit careful the resulting spin-network will have zero norm. And if the norm is zero, then the situation represented by the spin-network is not realizable (i.e. zero probability) according to the rules of quantum mechanics.

There are, in fact, two simple necessary requirements which must be satisfied at each vertex of a spin-network, for its norm to be non-zero. Notice first that all the spin-networks that I have explicitly drawn have the property that precisely *three* edges (i.e. units) come together at each vertex. (This isn't one of the 'requirements' I am referring to. It's just that I don't know how one could handle more general types of vertex within the scheme.) Suppose we have a vertex at which an a-unit, a b-unit and a c-unit come together (fig. 10).

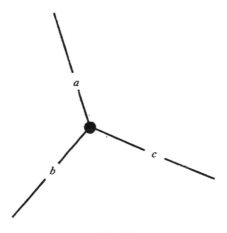

Fig. 10

Then for a spin-network containing this vertex to have a non-zero norm, it is necessary that the triangle inequality hold:

$$a+b+c \geqslant 2\max(a, b, c);$$

and furthermore that there be conservation of fermion number (mod 2):

$$a+b+c \quad \text{is even}.$$

These are, of course, properties that one would want to hold in real physical processes, with the interpretations that I have given to the spin-networks.

But even if these requirements hold at every vertex, the spin-network may still have zero norm. For example, each of the two types of spin-network shown in fig. 11 has zero norm, where $n \neq 0$ in the first case and $n \neq m$ in the second. In each case, the shaded portion represents some spin-network with no other free ends. In fact, the

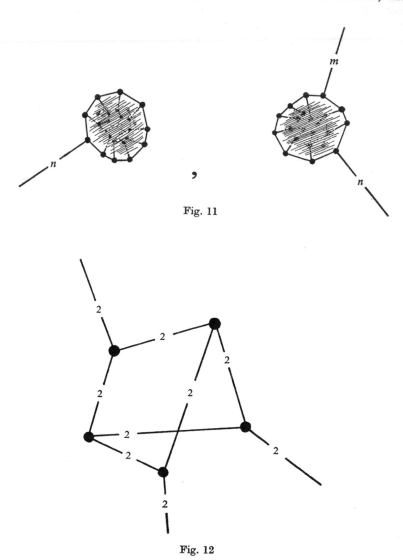

Fig. 11

Fig. 12

first is effectively a special case of the second, with $m = 0$. This is because any 0-unit can be omitted from a spin-network (if we also suitably delete the relevant vertices) without changing the norm. We may interpret the vanishing of the norm whenever $n \neq m$ in the second case as an expression of *conservation of total angular momentum*.

In addition to these cases, there are many particular spin-networks which turn out to have zero norm. One example is shown in fig. 12. But so far I have only been giving particular cases. Let us now pass to the general rule.

I shall give the definition of the norm in terms of a closely related concept, namely, what I shall call the *value* of a closed oriented spin-network. I call a spin-network *closed* if it has no free ends (e.g. analogous to a disconnected vacuum process). A spin-network which is not closed will not be assigned a value. The definition of *orientation* for a spin-network is a little difficult to give concisely. Any spin-network can be assigned two alternative orientations. Fixing the orientation of a closed spin-network will serve to define the *sign* of its value (which can be positive or negative). Roughly speaking the orientation assigns a cyclic order to the three units attached to each vertex—but if we reverse the cyclic order at any *even* number of vertices this is to leave the orientation unchanged. The orientation will change, on the other hand, if the cyclic order is reversed at an *odd* number of vertices.

I shall adopt the convention, when drawing spin-networks, that the orientation is to be fixed by the way that the spin-network is depicted on the plane. At each vertex we specify '*counter-clockwise*' as the cyclic order for the three units attached to the vertex. This defines the spin-network's orientation. The diagrams in fig. 13 illustrate an example of a closed spin-network with its two possible orientations.

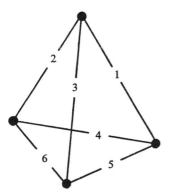

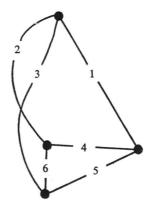

Fig. 13

It will also be convenient to use the representation of a spin-network as a drawing on a plane, in order to keep track of signs properly when defining the value. This may have the effect of making the definition *seem* less 'combinatorial' than it really is. Of course, the definition could be reformulated without the use of such a drawing if desired. Consider, then, a closed spin-network α depicted as a

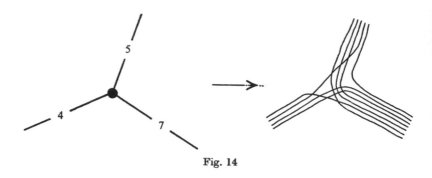

Fig. 14

drawing on a plane. Now, imagine each n-unit to be replaced in the drawing by n parallel strands. At each vertex, the strand ends must be connected together in pairs, but no two strands associated with the same n-unit are to be connected together. Let us call such a connection scheme a *vertex connection*. One such vertex connection is illustrated in fig. 14, while fig. 15 shows a non-allowable connection,

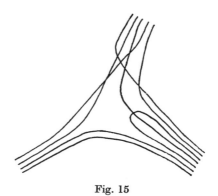

Fig. 15

since two strands of the 7-unit are connected to one another. The *sign* of a vertex connection is defined most simply as $(-1)^x$ where x is the number of intersection points between different strands at the vertex, as drawn on the plane. (These intersection points must be counted correctly if more than two strands cross at a point, or if two strands touch: and ignored if a strand crosses itself. It is simplest on the other hand, just to avoid such features by drawing the strands in general position and not allowing any strand connection to cross itself.) The sign of a vertex connection, in fact, does not depend on the details of how it is drawn, but only on the pairing off of the strands. The allowable vertex connection depicted above has -1 as its sign, since there are thirteen crossing points.

When the vertex connections have been completed at every vertex of a closed spin-network, then we shall have a number of *closed loops*, with no open-ended strands remaining. Consider, now, every possible way of allowably completing the vertex connection for the spin-network α. We form the expression

$$\text{value of } \alpha = \frac{\Sigma \pm (-2)^c}{\Pi n!}$$

where the summation extends over all possible completed allowable connection schemes, where the ' $\pm$ ' stands for the product of the signs of all the vertex connections, where c is the number of closed loops resulting from the vertex connections and where the product in the denominator ranges over all the units of the spin-network, n being the spin-number of the unit. The value of any closed spin-network always turns out to be an integer.

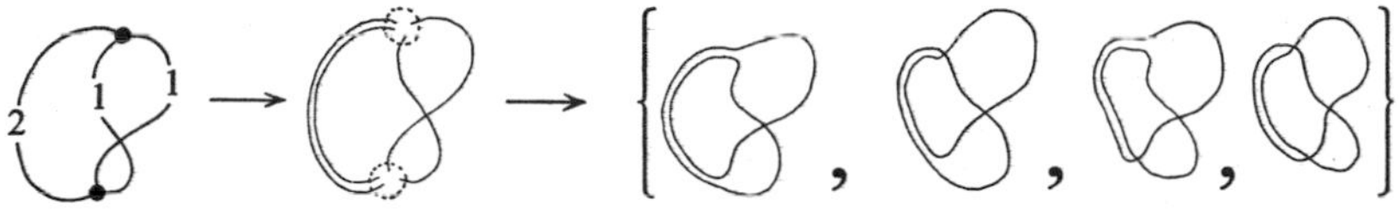

$$\rightarrow \text{value} = \frac{1}{2!1!1!} \{ + (-2) - (-2)^2 - (-2)^2 + (-2) \}$$

$$= -6.$$

Fig. 16

Let us consider a simple example, given in fig. 16. Note that the 'accidental' intersection, arising from the crossing of the two 1-units in the first drawing of the spin-network, does not contribute to the sign of the terms in the sum. Only the intersections at the vertex connections count.

The definition of the value of a closed oriented spin-network that I have just given is perhaps the simplest to state, but it is by no means the most useful to use in actual calculations. When the spin-networks become even slightly more complicated than the simple one evaluated above, the detailed calculations can become very unwieldy. A more useful procedure is to employ certain reduction formulae which can be used to express complicated networks in terms of simpler ones.† For this purpose, it will be necessary to introduce a slight variation of the spin-network theme; I shall consider the related concept of a *strand-network*.

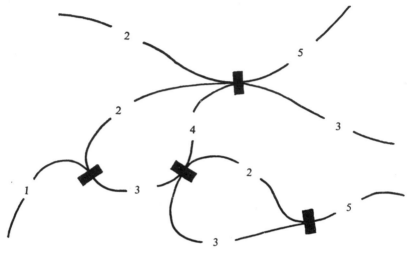

Fig. 17

A strand-network is a series of connections relating objects (which I shall still refer to as *n*-units) an example of which is depicted in fig. 17. The units are 'tied together' at various places, as indicated by the thick bar. Any spin-network can be translated into strand-network terms, by replacing each vertex according to the scheme shown in fig. 18. I thus introduce three more ('virtual') units at each vertex. A strand-network is *closed* if it has no free ends. Any closed (oriented) strand-network will have a *value* which is an integer (positive, negative or zero). This value will be chosen to agree with that defined for a spin-network, in the case of closed strand-networks obtained by means of the above replacement. Generally, to obtain the value of a closed strand-network β we employ the same formula as before:

$$\text{value of } \beta = \frac{\Sigma \pm (-2)^c}{\Pi n!}$$

† Diagrams closely related to spin-networks were introduced by Ord-Smith and Edmonds for the graphical treatment of quantum mechanical angular momentum. (See reference (1) for a detailed discussion.)

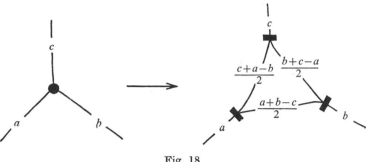

Fig. 18

where now the ' $\pm$ ' sign refers to the product of all the signs of all the permutations involved in each strand connection at which the strands come together. For example, one possible connection scheme for

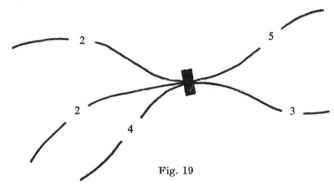

Fig. 19

'strand vertex' of fig. 19, would be that shown in fig. 20. This connection scheme would contribute a minus sign, since an odd permutation is involved. (There are nine crossing points—this is essentially

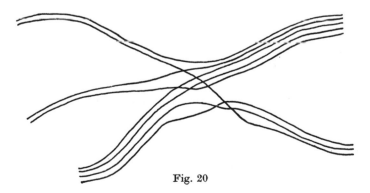

Fig. 20

the 'Aitken diagram' method of determining the sign of a permutation.) Notice that for a connection scheme to be possible at all, we require that the total of the spin-numbers entering at one side must equal the total of the spin-numbers leaving at the other. This *one* requirement now takes the place of the 'triangle inequality' and 'fermion conservation' that we had earlier.

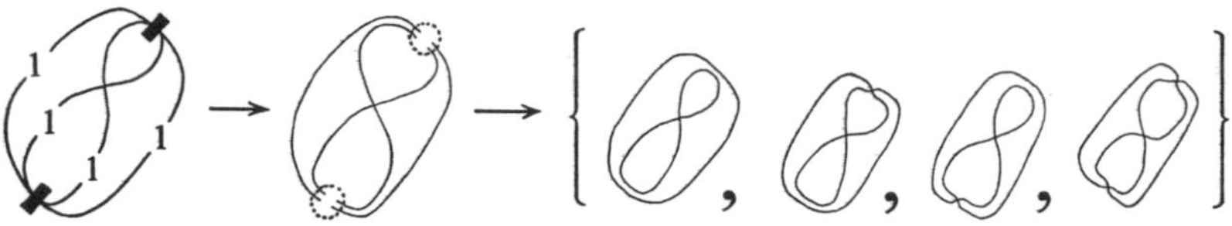

$$\therefore \text{ value} = \frac{1}{1\,!1\,!1\,!1\,!}\{+(-2)^2-(-2)-(-2)+(-2)\}$$

$$= 2.$$

Fig. 21

Let us evaluate the simple closed strand-network of fig. 21 as an example. Again there is an 'accidental' intersection depicted (where two 1-units cross) which does not contribute to the sign of the terms in the sum.

Let me list a number of relations and reduction formulae which are useful in evaluating strand-networks (fig. 22). (I am not going to prove anything here, but most of the relations are not hard to verify.) These relations may be substituted into any closed strand-network and a valid relation between values is obtained. Finally, let me make the remark that the value is *multiplicative*, that is to say, the value of the union of two *disjoint* strand-networks or spin-networks is equal to the product of their individual values.

I now come to the definition of the *norm* of a spin-network. A strand-network will likewise have a norm. This is simply obtained by drawing *two* copies of the spin-network (or strand-network), joining together the corresponding free end units to make a closed network, and then taking the modulus of the value of this resulting closed network. As an example, the norm of the spin-network consisting of a single vertex is found in fig. 23. An even simpler, example, depicted in fig. 24, is the norm of a single isolated n-unit.

Finally, let me describe how the norm may be used in the calculation of *probabilities* for spin-numbers, in the type of 'experiment' that we have been considering. (Again I shall give no proofs.) Suppose we start with a spin-network α, with an a-unit and a b-unit among its

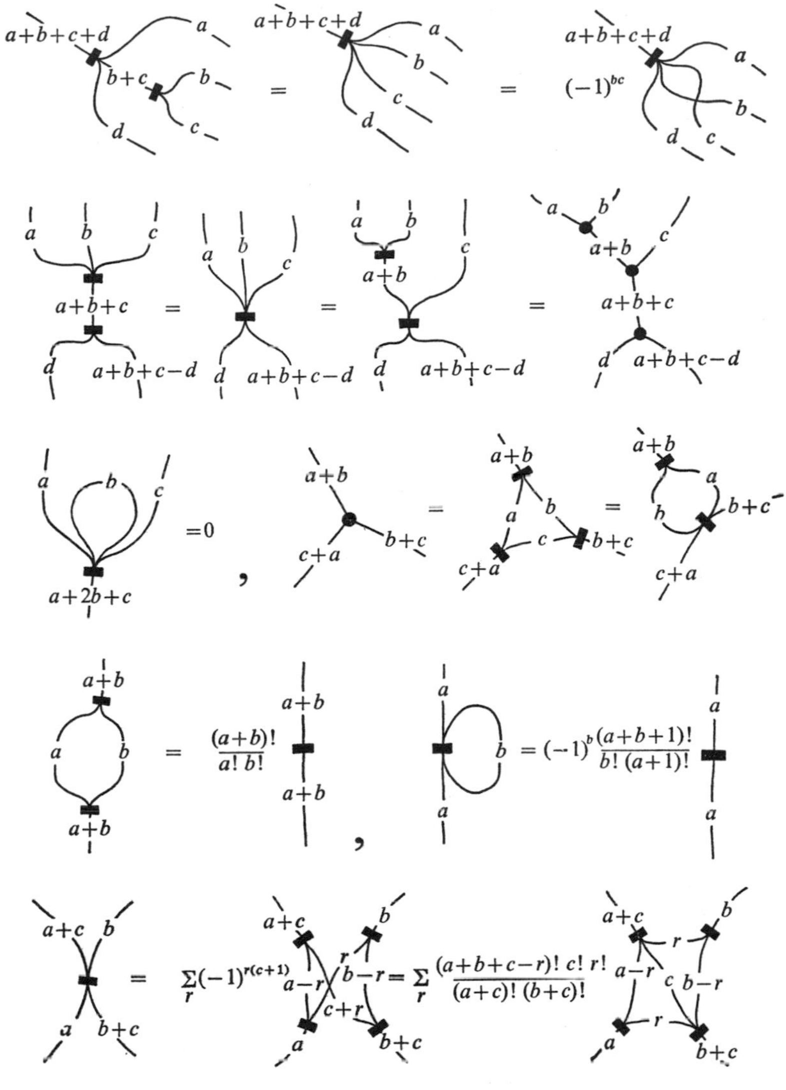

$$\frac{(a+b)!}{a!\,b!}$$

$$(-1)^b\frac{(a+b+1)!}{b!\,(a+1)!}$$

$$\sum_r (-1)^{r(c+1)} = \sum_r \frac{(a+b+c-r)!\,c!\,r!}{(a+c)!\,(b+c)!}$$

Fig. 22

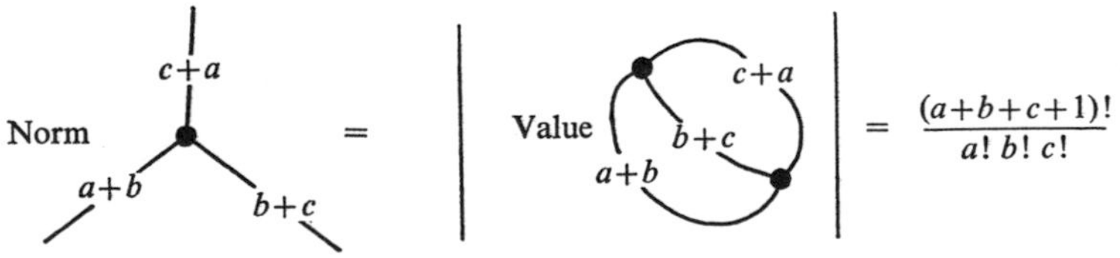

Norm $=$ Value $= \dfrac{(a+b+c+1)!}{a!\,b!\,c!}$

Fig. 23

free ends (fig. 25). Suppose the a-unit and the b-unit come together to form an x-unit, the resulting spin-network being denoted by β (see fig. 26). We wish to know (given α) what are the various probabilities for the different possible values of the spin-number x. Let γ denote the spin-network representing the coming together of the a-unit and

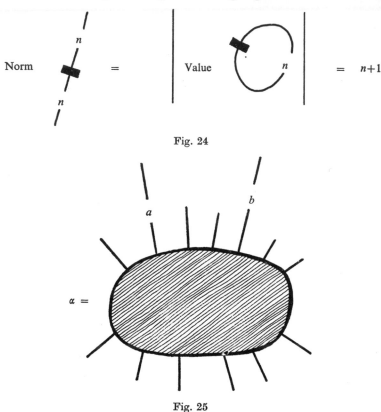

Fig. 24

Fig. 25

the b-unit to form the x-unit. Let ξ denote the spin-network consisting of the x-unit alone. These are illustrated in fig. 27. Then the required probability for the resulting spin-number to be x is

$$\text{probability} = \frac{\text{norm } \beta \text{ norm } \xi}{\text{norm } \alpha \text{ norm } \gamma}.$$

Using the explicit expressions for norm γ and norm ξ that were just given as examples (using a slightly different notation), we can rewrite this as

$$\text{probability} = \frac{\text{norm}\beta \; (x+1) \left\{\tfrac{1}{2}(a+b+x)+1\right\}!}{\text{norm}\alpha \left\{\tfrac{1}{2}(a+b-x)\right\}! \left\{\tfrac{1}{2}(b+x-a)\right\}! \left\{\tfrac{1}{2}(x+a-b)\right\}!}.$$

From the combinatorial nature of the definition of norm, it is clear that these probabilities must all be rational numbers. And with the interpretation of 'angle' that I have given, the three-dimensional Euclidean nature of the 'directions in space' that are obtained, is a *consequence* of these combinatorial probabilities.

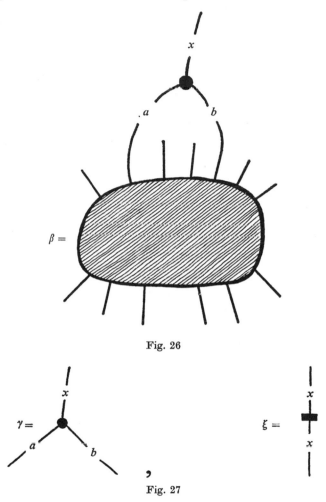

Fig. 26

Fig. 27

I should emphasize again that the space that I get out in the end is the one *defined* by the system itself and is not really the same space as the one that might have been introduced at the start if a conventional formalism had been used. Thus, although undoubtedly the reason that we end up with directions in a Euclidean three-dimensional space is intimately *related* to the fact that we start with representations of the rotation group SO(3), the precise logical connections are not at all clear to me. When I come to consider the generalization of all this to a relativistic scheme in a moment, this question will again present itself. I shall also need to consider the spatial locations of objects, not just their orientations. My model works with objects and the interrelations between objects. An object is thus 'located', either directionally or positionally in terms of its relations with other objects. One doesn't really need a space to begin with. The notion of space comes out as a *convenience* at the end.

Essentially, I have so far been using a non-relativistic scheme. The angular momentum is not relativistic angular momentum. From a four-dimensional viewpoint, the directions I get are those ortho-gonal to a given timelike direction, i.e. directions in three-dimensional space. All the particles are going along in this same timelike direction. Perhaps they can knock each other a little bit, but they are not really moving very much. They just transfer angular momentum back-wards and forwards. All the particles are, strictly speaking in the same place, not moving relative to one another. Consequently, one does not have any problem of mixing between orbital and spin angular momentum. Once I allow orbital contributions to come in, then I must drastically change the scheme, since now not only is the question of 'direction' and 'angle' involved, but so also is 'position' and 'distance'. Thus, if one thinks of *real* particles moving relative to each other, then there is the problem not only of doing things relativistically, but also of bringing in actual displacements between particles. Consider two particles in relative motion. Suppose they come together and combine to form a system with a well-defined total spin. Then to obtain the spin of the combined system, we cannot just add up the individual spins because we have to bring in the orbital component. There is a mixture of the actual displace-ments in space with the angular momentum concept. So I spent a long time thinking how one should combine rotations and displace-ments together into an appropriate relativistic scheme. Eventually I was led to consider a certain algebra for space-time which treats linear displacements on the same footing as it treats rotations. Thus, linear momentum is treated on a similar footing to angular momentum.

Now you might raise the objection that linear momentum has a continuous spectrum, while it is only for angular momentum that one has discreteness. This is a problem of some significance to us. My answer to it is roughly the following: each particle has its own discrete spectrum for its angular momentum. When two particles are considered together as a unit, then again there is a discrete spectrum for the combined system. The way these 'spins' add up implicitly brings in the relative motion between the two particles. So the momentum is brought in through the back door, in a sense, for one could be always talking in terms of 'bound systems'.

I consider momentum states as being linear combinations of angular momentum states. There is indeed a problem to see how this continuous momentum should be built up from something discrete, but, in principle, there is nothing against it. In effect, the idea is that the momentum should be brought in indirectly. I would propose that, in a sense, there should *not* be a well-defined distinction between momentum and angular momentum—except in the limit. Individual particles and simple systems would not really 'know' what momentum is. Like the idea of 'direction' that I considered earlier, it would be only in the limit of large systems that the concept of momentum really attains a well-defined meaning. Smaller systems might retain a combined concept of momentum and angular momentum, but these things would only sort themselves out properly in the limit.

The algebra I have used to treat linear displacements and rotations together, or linear and angular momentum together, I call the algebra of *twistors*. I have used the term 'twistor' to denote a 'spinor' for the six-dimensional $(+ + - - - -)$ pseudo-orthogonal group $O(2, 4)$. The twistor group is the $(+ + - -)$ pseudo-unitary group $SU(2, 2)$, which is locally isomorphic with $O(2, 4)$. In turn, $O(2, 4)$ is locally isomorphic with the fifteen-parameter (local) conformal group of space-time. Under a conformal transformation of space-time, the twistors will transform (linearly) according to a representation of the group $SU(2, 2)$.

The basic twistor is a four-complex-dimensional object. We can thus describe it by means of four complex components Z^α:

$$(Z^\mu) = (Z^0, Z^1, Z^2, Z^3).$$

The complex conjugate of the twistor Z^α is an object $\bar{Z}_\alpha$ whose components
$$(\bar{Z}_\alpha) = (\bar{Z}_0, \bar{Z}_1, \bar{Z}_2, \bar{Z}_3)$$
are given (according to a convenient co-ordinate system) by

$$\bar{Z}_0 = \overline{(Z^2)}, \quad \bar{Z}_1 = \overline{(Z^3)}, \quad \bar{Z}_2 = \overline{(Z^0)}, \quad \bar{Z}_3 = \overline{(Z^1)}.$$

This implies that the Hermitian form $Z^\alpha \bar{Z}_\alpha$ (summation convention assumed) has signature $(+ + - -)$, which is required in order to give $SU(2, 2)$. (I have already described these objects [2] and their geometrical significance in Minkowski space-time, and a later paper [3] goes into some further developments, including some of the quite surprising aspects of the theory which arise when one starts to describe physical fields in twistor terms.)

When $Z^\alpha \bar{Z}_\alpha = 0$, I call Z^α a *null* twistor. A null twistor Z^α has a very direct geometrical interpretation in space-time terms. In fact, Z^α defines a *null straight line*, which we can think of as the world line of a zero rest-mass particle. (An important aspect of twistor theory is that *zero rest-mass* is to be regarded as more fundamental than finite rest-mass. Finite mass particles are viewed as composite systems, the mass arising from interactions.) The twistor λZ^α, for any non-zero complex number λ, defines the same null line as does Z^α. But we can distinguish Z^α from λZ^α by assigning to the particle a 4-momentum (in its direction of motion) and, in addition, a sort of 'polarization' direction (both constant along the world line of the particle). When Z^α is replaced by rZ^α (r real) the momentum is multiplied by r^2. When Z^α is replaced by $e^{i\theta}Z^\alpha$ (θ real) the 'polarization' plane is rotated through an angle 2θ. If Y^α is another null twistor, the condition for the null lines represented by Y^α and Z^α to *meet* is

$$Y^\alpha \bar{Z}_\alpha = 0,$$

i.e. this is the condition for the two particles to 'collide'. (To be strictly accurate we have to include the possibility that they may 'meet at infinity'. In addition, some of the null twistors describe 'null lines at infinity' rather than actual null lines.)

The *non-null* twistors are divided into two classes according as $Z^\alpha \bar{Z}_\alpha$ is positive or negative. If $Z^\alpha \bar{Z}_\alpha > 0$, I call Z^α right-handed; if $Z^\alpha \bar{Z}_\alpha < 0$, left-handed. In Minkowski space-time, one can give an interpretation of a non-null twistor in terms of a twisting system of null lines. The helicity of the twist is defined by the sign of $Z^\alpha \bar{Z}_\alpha$. In more physical terms, the twistor Z^α (up the phase) describes the momentum and angular momentum structure of a zero rest-mass particle.[†] We can make the interpretation that $Z^\alpha \bar{Z}_\alpha$ is (twice) the *intrinsic spin* of the particle, measured in suitable units, with a *sign* defining the helicity. If $Z^\alpha \bar{Z}_\alpha \neq 0$, then it is not possible actually to localize the particle as a null straight line. Only if $Z^\alpha \bar{Z}_\alpha = 0$ do we

† Using a convenient co-ordinate system, we can relate the momentum P_a and the angular momentum M^{ab} ($= -M^{ba}$) to the twistor variables Z^α, $\bar{Z}_\alpha$ by:

$$P_0 + P_1 = 2^{\frac{1}{2}} Z^0 \bar{Z}_2, \quad P_0 - P_1 = 2^{\frac{1}{2}} Z^1 \bar{Z}_3, \quad P_2 + iP_3 = 2^{\frac{1}{2}} Z^1 \bar{Z}_2,$$
$$P_2 - iP_3 = 2^{\frac{1}{2}} Z^0 \bar{Z}_3, \quad M^{23} + iM^{01} = Z^1 \bar{Z}_1 - Z^0 \bar{Z}_0, \quad M^{23} - iM^{01} = Z^3 \bar{Z}_3 - Z^2 \bar{Z}_2,$$
$$M^{13} + M^{03} + iM^{02} + iM^{12} = 2Z^2 \bar{Z}_3, \quad M^{13} - M^{03} + iM^{02} - iM^{12} = 2Z^3 \bar{Z}_2,$$
$$M^{13} + M^{03} - iM^{02} - iM^{12} = 2Z^1 \bar{Z}_0, \quad M^{13} - M^{03} - iM^{02} + iM^{12} = 2Z^0 \bar{Z}_1.$$

get a uniquely defined null line which we can think of as the world line of the particle; otherwise the particle to some extent spreads itself throughout space.

The twistor co-ordinates Z^0, Z^1, Z^2, Z^3, together with their complex conjugates $\bar{Z}_0$, $\bar{Z}_1$, $\bar{Z}_2$, $\bar{Z}_3$, can be used in place of the usual x, y, z, t and their canonical conjugates p_x, p_y, p_z, E. In fact, anything that can be written in normal Minkowski space terms can be rewritten in terms of twistors. However, in principle, the twistor expressions for even quite simple physical processes may turn out to be very complicated. But in fact it emerges that the basic elementary processes that one requires, can actually be expressed very simply if one goes about it in the right way. Analytic (holomorphic) functions in the Z^α variables play a key role. So does contour integration.

We can regard the Z^α as quantum operators under suitable circumstances. Then $\bar{Z}_\alpha$ can be regarded as the canonical conjugate of Z^α. (I shall go into the reasons for this a little more later.) We have commutation rules

$$Z^\alpha \bar{Z}_\beta - \bar{Z}_\beta Z^\alpha = \delta^\alpha_\beta \hbar.$$

Then, since Z^α and $\bar{Z}_\alpha$ do not commute, we must re-interpret the expression for the *spin-helicity* $\frac{1}{2}n$ as the *symmetrized* quantity,

$$\tfrac{1}{4}(Z^\alpha \bar{Z}_\alpha + \bar{Z}_\alpha Z^\alpha) = \tfrac{1}{2}n\hbar.$$

Only zero rest-mass states can be eigenstates of this operator. The eigenvalues of n are $\ldots - 2$, -1, 0, 1, 2, $\ldots$ The operators for the ten components of momentum and angular momentum (together with those for the five extra components arising from the conformal invariance of zero rest-mass fields) are

$$Z^\alpha \bar{Z}_\beta - \tfrac{1}{4}\delta^\alpha_\beta Z^\gamma \bar{Z}_\gamma$$

in twistor notation. The usual commutation rules for momentum and angular momentum are then a consequence of the twistor commutation rules.

One idea would now be to use this fact and simply let the twistors take the place of the *two-component spinors* that lay 'behind the scenes' in my previous non-relativistic approach, and then to attempt to build a concept of a four-dimensional space-time from whatever graphical algebra arises from the twistor rules. I have not attempted to do quite this, as yet, since I am not sure that it is exactly the right thing to do. There are certain other aspects of twistor theory which should really be taken into account first.

Let me mention one particular point. It is a rather remarkable one. If the twistor approach is going to have any fundamental significance in physical theory, then it ought, in principle at least, to be possible to carry the formalism over and apply it to a *curved* space-time, rather than just a flat space-time. These objects, as originally defined, are very much tied up with the Minkowski flat space-time concept. How can we carry them over into a curved space-time? Actually, a twistor for which $Z^\alpha \bar{Z}_\alpha = 0$ carries over very well. Its interpretation is now simply as a *null geodesic* (i.e. world line of a freely moving massless particle) with a momentum (pointing along the world line) and a 'polarization' direction (both covariantly constant along the world line). On the other hand, it does not seem to be possible to interpret a *non-null* twistor, in a general curved space-time, in precise classical space-time terms. Nevertheless it turns out to be convenient to *postulate* the existence of these non-null twistors—as objects with no classical realization in curved space-time terms. (In a sense, twistors are more appropriate to the treatment of *quantized* gravitation† than of classical general relativity.)

Let us concentrate attention, for the moment, on the *null* twistors only so that we can consider purely geometrical questions. We are interested in properties of null geodesics which refer to each geodesic as a *whole* and not to the neighbourhood of some point on a geodesic. Consider, for example, the condition of orthogonality between twistors. We have seen that in flat space-time, the condition of orthogonality $Y^\alpha \bar{Z}_\alpha = 0$ between two twistors Y^α, Z^α corresponds to the *meeting* of the corresponding null lines. In curved space-time this is not really satisfactory, because although I can tell whether two null geodesics are going to meet if I look in the neighbourhood of the intersection point, if I look somewhere else, I can't see whether or not they will meet, because the curvature may have bent them away from each other. So, in fact, the orthogonality property is not something which is preserved, as an invariant concept, when one turns to curved space-time. On the other hand, certain things *are*

---

† Since this lecture was delivered, there have been some developments in twistor theory of relevance to this discussion. It seems to be possible to express quantized gravitational theory in twistor form. By means of $3k$-dimensional contour integrals in spaces of many twistor variables, one can apparently calculate scattering amplitudes for processes involving gravitons, photons and other particles. Diagrams arise which can be used to replace the spin-networks of the formalism described here. It is not impossible that the calculations can be reduced to a set of comparatively simple combinatorial rules, but it is unclear, as yet, whether this is so. The work is still very much at a preliminary stage of development and many queries remain unanswered.

preserved; and, somewhat surprisingly, they correspond to assigning a *symplectic* structure to the twistor space.

This symplectic structure is expressed (in appropriate co-ordinates) as the invariance of the 2-form

$$dZ^\alpha \wedge d\bar{Z}_\alpha$$

(using Cartan notation). Strictly speaking, this requires the postulated non-null twistors, in addition to the null ones. The null twistors only form a seven-real-dimensional manifold, whereas a symplectic manifold must be even-dimensional. The null twistor manifold must be embedded in the eight-real-dimensional manifold consisting of *all* twistors. The structure of the null twistor manifold is that induced by the embedding in this eight-dimensional symplectic manifold. In addition to the symplectic structure (and closely related to it), the expressions

$$Z^\alpha \bar{Z}_\alpha, \quad Z^\alpha d\bar{Z}_\alpha, \quad Z^\alpha \frac{\partial}{\partial Z^\alpha}$$

are also invariant. All these invariant quantities can be interpreted, to some extent, in terms of the geometrical properties of null geodesics. But it will not be worthwhile for me to go into all this here.

The invariance of the symplectic structure of the twistor space for curved space-time can be re-expressed as the invariance of the *Poisson brackets*

$$[\psi, \chi] = i \frac{\partial \psi}{\partial Z^\alpha} \frac{\partial \chi}{\partial \bar{Z}_\alpha} - i \frac{\partial \chi}{\partial Z^\alpha} \frac{\partial \psi}{\partial \bar{Z}_\alpha}.$$

This strongly suggests that in the passage to quantum theory, $\bar{Z}_\alpha$ should be regarded as the conjugate variable to Z^α. Thus, we are led to the commutation rules I mentioned earlier, relating quantum operators Z^α and $\bar{Z}_\alpha$. These commutation rules in turn give us the commutation relations for momentum and angular momentum, as I indicated before. This suggests that there may possibly be some deep connection between these commutation rules (or perhaps some slight modifications of them) and the curvature of space-time.

The picture that one gets is that in some sense the curvature of space-time is to do with canonical transformations between the twistor variables Z^α, $\bar{Z}_\alpha$. Suppose we start in some region of space-time where things are essentially flat. Then we can interpret Z^α and $\bar{Z}_\alpha$ in a straightforward way in terms of geometry and angular momentum, etc. Suppose we then pass through a region of curvature to another region where things are again essentially flat. We then find that our interpretations have undergone a 'shift' corresponding to a canonical transformation between Z^α and $\bar{Z}_\alpha$. In effect the 'twistor position' (i.e. Z^α) and 'twistor momentum' (i.e. $\bar{Z}_\alpha$) have got mixed up. Somehow it is this mixing up of the 'twistor position' and 'twistor momentum' which corresponds to what we see as space-time curvature.

Going back to my original combinatorial approach for building space up from angular momentum, we can ask now whether such a combinatorial scheme could be applied to the twistors. Might it be that, instead of ending up with a flat space, we could end up with a curved space-time? Even if I start with the commutation rules appropriate just to the Poincaré group, or perhaps the conformal group, it is obvious that I must end up with essentially the same space that I 'start' with? One has to define the things with which one builds up geometry (e.g. points, angles, etc.), in *terms* of the physical objects under consideration. It is not at all clear to me that the geometry that is built up in one region will not be 'shifted' with respect to the geometry built up in some other region. Is it then not possible that a space-time possessing curvature might be the result? That is really the final point I wanted to make.

REFERENCES

(1) Yutsis, A. P., Levinson, I. B. and Vanagas, V. V. *Mathematical apparatus of the theory of angular momentum* (Jerusalem: I.P.S.T., 1962).
(2) Penrose, R. *J. Math. Phys.* (1967) **8**, 345.
(3) Penrose, R. *Int. Jl Theor. Phys.* (1968) **1**, 61.

A NOTE ON DISCRETENESS, PHASE SPACE AND COHOMOLOGY THEORY

B. J. HILEY[†]

This paper originates in an attempt I made at the colloquium to provide an introduction for Atkin's talk, a modified form of which follows this paper. I originally set out from the natural intuitive idea of dividing phase space into cells to introduce quantisation and showed that a more radical mathematical approach than the simple mechanical one suggested by Ishiwara [1] was necessary. This, Atkin was to supply in more detail. The ideas in this paper stem from a suggestion made to me by Bohm as early as 1963. He pointed out that the work of Hodge, [2] which uses de Rham cohomology theory to discuss Maxwell's equations, implicitly contained certain features which would be necessary to go beyond quantum mechanics. In particular the use of the incidence matrices suggested a natural way of introducing discreteness. Atkin's method is different and depends on his own idea of 'scales'.

My motivation for attempting to go beyond quantum theory springs mainly from the problems that arise in high energy phenomena (i.e. elementary particle physics) which, I believe, have their roots in difficulties presented by the quantum description. I do not question the validity of the quantum theory in the low energy domain (i.e. where there is a fixed number of slowly moving particles), although even here, it should be emphasised that the empirical success of the quantum theory does not necessarily imply the non-existence of valid alternatives. [3]

These difficulties mentioned above cannot be overcome by simply modifying one or two axioms of the present theory. Nor will they be overcome by a return to classical concepts. What is called for is a more radical change of viewpoint which is no longer based on continuous descriptions in space-time and which therefore raises new questions of a kind that are not considered in the present theories.

The space-time continuum, together with the associated notions of particle, trajectory, field, etc., have played a dominant role in classical physics. Both the particle and the field require *local dynamical*

† The author would like to thank D. Bohm, R. H. Atkin and D. Schumacher for their helpful discussions.

descriptions (using differential equations) which give a completely deterministic account of the universe. All *aspects* of a system can be observed *simultaneously* and any interaction during the observational process can be made effectively negligible; or, if this is not possible in practice, it can be allowed for in the final calculation.

The quantum theory necessitated a deep and radical change from the above ideas. It was no longer possible to observe *all aspects* of the system *simultaneously*. Indeed, the quantum of action brought into question the validity of the separation of a system from the observing apparatus. The outcome of any experiment was found to depend upon the form of observation and therefore the description required a specification of the apparatus and the general conditions of the experiment as a whole. In this type of description the precise definition of the particle trajectory has no place. The emphasis is on the *wholeness of the description*, that is, the form of the experimental conditions is a whole with the content of the experimental outcome and in which analysis of the disjunction between object and apparatus is not relevant. This is in contrast to the *local* nature of the classical description in which a detailed analysis is not only relevant, but also necessary. In this context wholeness is an informal notion and can be taken into account formally with the aid of the wave-function, while the incompatibility of different kinds of observations is formally taken into account by means of the non-commutation of operators.

These notions are put on a firm mathematical foundation by introducing the Hilbert space formalism. However, this mathematical formulation requires that we specify the relationship of operators to the results of observations. In order to determine this relationship, it has been assumed that it is necessary to start from classical concepts, i.e. to consider charged particles with a given energy-momentum at given points in space-time interacting through electromagnetic fields which are specified at every point in space-time, etc. These are then *translated* into the quantum formalism by a set of rules. However, the very notion of translation or interpretation implies a disjunction and it is this disjunction which is being denied by the informal ideas of quantum theory. This disjunction implies the possibility of analysis which, as we have seen, is not relevant in this context.

In general, physicists tend to regard this type of informal consistency argument as not very significant because it is believed that the essence of the theory lies in the mathematical formalism, which is assumed to be independent of the way we talk about the experiment.

However, the way we talk informally is just as significant as the formalism.

Indeed, the role of informal language in classical and in quantum theory has been discussed recently [4] and it has been made clear that fundamental changes of theory require radical changes, not only in the mathematics, but also in the ordinary language of description.

Although it may be convenient for some purposes to distinguish between the informal and the formal languages, they actually constitute a whole in which informal notions are inseparably intertwined with the formalism. Because the quantum theory introduces into physics a new kind of informal description, it is now necessary to consider very carefully the interplay between the informal and the formal. A disjunction or even a lack of harmony between these two aspects of the language will tend to lead to a situation in which it is impossible to know what to do next when unfamiliar ground is encountered. Any attempt to remove these difficulties must necessitate a change both in the informal and formal languages. Of course it is not possible nor, indeed, is it necessary to *analyse* these changes and their interrelation in any depth, but it is desirable to be aware of the possibility of change. If the subtler changes in the informal language are not noticed then attempts to put the new ideas into the old in- formal language will lead to contradictions and eventually to confusion.

The lack of harmony that I am referring to in the general arguments used above can be seen from a more particular point of view. Experiments indicate that it is necessary to introduce ideas like energy-levels, quanta, space-quantisation, quantisation of vortices, etc. into our theories. That is, an inherent discreteness is needed in our descriptions. Yet our present theories use a continuous space-time backcloth, together with other classical concepts such as particle, mass, charge, etc. The result is that there are endless arguments about 'quantum jumps' and about the type of questions raised by the Einstein–Podolsky–Rosen paradox.[5] We will see that in developing a language in which the essential informal form is wholeness, rather than disjunction, discreteness arises quite naturally.

Before it is possible to develop such a language, it is necessary to examine briefly the notion of a particle. The very concept of a particle implies the existence of a disjoint object with a definite position and definite momentum at every moment of time, and this in turn implies the existence of particle trajectories in a classical space-time continuum. However, we have already seen that this notion is no longer basic in the quantum theory. Indeed, to Bohr [6] the quantum of action meant that it was no longer possible to have unambiguous descriptions of microscopic phenomena, and he was thus led to propose the principle of complementarity which, when applied to the particular case of dynamics, implies descriptions that can be *either* space-time descriptions *or* causal descriptions, but never *both* together.

These considerations have been recognised almost since the beginning of the quantum theory but the particle idea has been retained mainly through the use of the classical informal language. For example, physicists are very used to the idea of a photon and think of it very much as if it were a particle even though one can *never* specify its position in space-time.

I would like to suggest that in any new theory the retention of the notion of a particle as a basic informal notion may be *irrelevant* and misleading particularly in interpreting high energy phenomena. Do the tracks that are seen in the bubble chambers, photo-emulsions, etc., imply the existence of a material object? If we look closely at the tracks we see a series of dots which, when joined point by point, give a very irregular curve. To interpret the track as a particle track, the first step has been to *replace* the set of dots by a smooth curve which is obtained by using some convenient averaging technique over the dots. The continuous curve, therefore, represents an ensemble average both in the sense that each curve is an abstraction obtained by averaging over a collection of dots, and in the sense that many tracks are generally averaged in order to determine a typical experimental result. The particle properties arise only as a result of an ensemble average on many individual observations and therefore the particle and its trajectory is only relevant in a certain rather abstract context, the precise nature of which must be made clear by the new theory.

It is true that in certain formal languages, such as field theories and *S*-matrix theories, the particle concept has been modified considerably, but at present it is not clear in exactly what way it has been changed. In consequence, there is a degree of confusion when interpreting the formalism.

In most field theories, the informal language still needs the particle concept since the algorithm requires the consideration of two free particles which are then brought into interaction. If the interaction is weak, the perturbation theory can be used; if the interaction is strong, then this cannot be done. In fact, there are interactions which are so strong that it would make no sense to try, *even informally*, to think that the objects have separate existence. In view of all these criticisms, therefore, let us seriously consider the suggestion that the particle should *not* be regarded as a basic informal notion.

Furthermore, the notion of a field used as a basic notion must also be called into question. A field requires the definition of a numerical value at every point in space-time and this definition contains explicitly both the notions of locality and separateness. However, this is not in harmony with the quantum notion of wholeness. Hence the notion of a field cannot be regarded as a basic notion and must also be dropped. In fact, we can no longer use any description containing the continuum notion, either explicitly or implicitly, since this notion always leads to the possibility of subdivision into parts. This means, of course, that all descriptions using continuous co-ordinates are irrelevant and must be dropped. Thus the starting point of all classical dynamical descriptions is denied us and we are faced with the question of where to start.

Let us first consider classical phase space. This is locally isomorphic to a Euclidean space of the same dimensionality and, as such, has a simple structure. The experiments that led to quantum theory provide strong indications however that phase space has a much richer—in fact, a cellular—structure. Theories of cell structure in phase space received some attention a considerable time before the Hilbert space formalism of quantum theory was developed, but has since been neglected. Apart from this early work, quantum theory itself furnishes evidence for such a structure. For example, the uncertainty principle suggests a structure which is in some way cellular in the sense that the position of a particle in phase space can only be specified to within a volume h^3. Second quantisation procedures also suggest a cellular structure in phase space in the sense that each cell corresponds to what is now called the state of a system. Further evidence for cellular structure has been pointed out by Aharonov [7] who considered the possibility of finding functions $f(q)$ and $g(p)$ such that $[f(q), g(p)] = 0$. It is not difficult to show that pairs of functions which satisfy this condition do exist. The functions are periodic, with periods inversely proportional to each other.

This cellular structure of phase space has generally been regarded as a *restriction* which must be imposed on a continuous description. In other words, the cellular structure has been regarded as a *limitation* on the simultaneous specification of position and momentum. However, I suggest that the cells must be regarded as *fundamental* and as such introduce novel qualitative features into the description. In consequence the use of co-ordinate descriptions are inadequate to deal with the more subtle structures which may be necessary for the description of physics, particularly in the microdomain.

Let us now consider in what way the cellular structure introduces the essential feature of wholeness into the description. To do this let us first assume that phase space is a Euclidean continuum. In this case the uncertainty principle only specifies the volume of the cells in phase space while the *shape* is determined in some sense by the apparatus and the set of experimental conditions used for a particular experiment. This view can be illustrated by considering the Heisenberg microscope experiment. It is not difficult to see that in this case the shape of the cell depends upon the energy of the incident radiation and the nature of the objective of the microscope. If the wavelength of the incident radiation is increased for the same microscope objective then the cells become elongated in the x-direction with a corresponding decrease in the p_x-direction, etc.

In his original discussion, Heisenberg used the traditional method of analysing the process into parts (e.g. the incident photons, the particle, the microscope, etc.) and this led to the idea that there was a 'particle which was being disturbed'. Further, by emphasising the constancy of $\hbar$ in the relation $\Delta x \Delta p_x \sim \hbar$ one tends to be led to the notion that the 'disturbance' is dependent only on the *size* of the cell in phase space. In this way the overall experimental conditions were tacitly dismissed as irrelevant. Thus one is left with the impression that the wholeness of the description is unnecessary. However, this is not true for, as we have seen, it is the *shape* of the cell which is the description that is associated to the experimental conditions and to neglect this shape is to miss the essential novelty introduced by the quantum of action.

In the quantum description, the shape of the cell in phase space is now inseparably amalgamated with the experimental conditions. The overall situation is characterised by a kind of wholeness, in which analysis (into cells plus experimental conditions) is not relevant. This situation is still the same when the cells are not embedded in a continuum which means that the cells should not be regarded as corresponding to disjoint and independently existing *things* in phase space (they are basically different from objects in space-time which can be described by co-ordinates). Nevertheless, there is still a certain similarity between co-ordinates and cells. The cells, however, are to be regarded as a way of describing structures that are *not* orders of point particles, point events, etc., but rather structures associated with the experimental conditions. Our task is to develop a suitable mathematical language using cellular structure as a basic description.

Mathematical descriptions using cells rather than co-ordinates are well known in homology and cohomology theory [8] and I shall use some of the descriptive terms of these theories to describe cells in phase space. As we will see, the notion of discreteness arises quite naturally in these theories. However, great care must be exercised in using these descriptions since they were not originally developed to discuss the problems that we are discussing.

In homology theory one starts with cells called simplexes as primitive and fundamental terms. These include the 1-simplex (which generalises the ordinary notion of a line), the 2-simplex (generalising the triangle), the 3-simplex (generalising the tetrahedron) etc. In the early phases of development of the theory, the various simplexes were regarded as covering a differential manifold. Continuous curves, surfaces, etc., were covered by chains of the appropriate dimension, while closed curves, surfaces, etc. were covered by corresponding cycles.

Although this development gives an intuitive meaning to the simplexes, chains and cycles, such a procedure is not necessary for their definition. Instead, it is possible to introduce the notion of an *abstract* simplex, not related to any manifold. A collection of these abstract simplexes is called a complex.

For our purposes the important structures on an abstract complex can be discussed in terms of Abelian groups. Each group is generated by the appropriate simplexes with integer coefficients. The chain group contains a sub-group of cycles, and mappings from the cycles into the integers generate the group of cocycles. The group of cycles is said to form an homology and the group of cocycles a cohomology.

Let us consider a particular example which illustrates how cohomology theory can provide a discrete description of a simple cubic lattice structure containing only edge dislocations. We initially suppose that the lattice is embedded in a R^3. If the plane density is described by $p_\mu(x)$ and the unit crystallographic vectors by $e_i(x)$, then the number of crystal planes crossed by each lattice vector is $n_i = \sum_\mu p_\mu e_i^\mu$

where n_i is an integer. In a set of lattice vectors which formed a closed loop, Γ, the net number of planes crossed is the algebraic sum of n_i over each of the given set of lattice vectors in Γ. The resulting integer, N, will evidently be the number of edge dislocations contained in Γ. Another way of writing this would be $N = \oint_\Gamma p_\mu dx^\mu$.

In terms of cohomology theory the lattice is described by a complex of interpenetrating lines and planes. In this complex the crystal vectors are regarded as 1-chains and the planes as 2-chains. The sum of n_i over Γ is then called the value of a 1-cochain on a 1-cycle. In this way the lattice structure with its dislocations defines an abstract complex with a cohomology.

I want to suggest that the use of cohomology theory goes far deeper than the above example suggests. It is well known that Maxwell's equations can be written in terms of differential forms which are used in de Rham cohomology theory. Misner and Wheeler [9] have already shown that in this language, electric charge is like a 'wormhole' which traps the electromagnetic field in the space. However, the de Rham theory is associated with an exterior derivative on a triangulated manifold and therefore relies on our informal notions of the continuum. What we wish to suggest is that the continuum is not necessary and that Maxwell's equations, in fact, describe a discrete structure (with cell-like properties) which possesses a boundary operator (corresponding to the exterior derivative) and in consequence a cohomology. Similarly the Hamilton–Jacobi, Schrödinger and Dirac equations can be expressed in terms of suitable coboundary operators which implicitly define corresponding cohomologies. These points are discussed in further detail in Atkin's paper.

Let us now turn to consider in what way the new description based on the simplicial complex will help us to understand high energy phenomena. Remember that a particle is to be treated as a secondary concept which arises from some invariant feature of the phase space structure which exhibits particle-like properties. The easiest way to illustrate what I have in mind is to consider a limiting case of a simplicial complex which has a particularly simple structure but which will, nonetheless, bring out some of the main ideas. Consider again the simple cubic lattice structure. In the region of a simple edge dislocation, the crystal structure is changed both topologically and metrically. As the edge dislocation migrates through the crystal the topological feature or singularity moves with it, giving the impression that the particle is moving through the background of the lattice. The dislocation can even have an inertial mass associated with it. [10] The metrical distortion can be interpreted as a field accompanying the 'particle' although I do not want to suggest that the 'particle' only exists at the singularity. The 'particle' is the *whole* structure. There is a sufficient variety of dislocations to accommodate the known types of 'elementary particles' and it is also possible to use the analogy to illustrate the processes of creation and destruction.[11] Although the analogy can be taken further, I believe this structure is far too simple to be of much use in the case of the microdomain. Furthermore, the experimental conditions are not explicitly included in the description. However, the example does illustrate the type of idea involved.

A rather interesting result can be obtained without the necessity of considering the structure in detail. If the metrical distortion in the neighbourhood of a singularity is regarded as being analogous to a 'bump' or small region of intense curvature in a continuous manifold representation of phase space, then this bump can be regarded as the particle. If the symplectic structure of phase space is carried through into the new description, then any localised structure can be smeared out over the entire space under the transformations of the symplectic group: the original localised structure and the smeared out structure will be dynamically equivalent. However, particle-like properties appear from localised structures and it is possible to prevent structures becoming smeared out by introducing a simple stability criterion which immediately leads to a space with unitary symmetry. By use of a description closely related to a Yang–Mills field type of argument, it is possible to obtain a current algebra of the same type that is being used in present theories of high energy phenomena. This description enables the currents to be interpreted geometrically in the continuous limit. However, what is important from our point of view is that the currents can be more easily interpreted in terms of the invariants arising from the description based on the simplicial complex. A detailed account of this work will be published elsewhere.

REFERENCES

(1) Ishiwara, J. — *Tokyo Sugaku Bulungakkawi Kizi* (1915) **8**, 106.
(2) Hodge, W. V. D. *The Theory and Applications of Harmonic Integrals* (Cambridge University Press, 1952).
(3) Feyerabend, P. K. *Frontiers of Science and Philosophy* (Allen and Unwin, 1964), 189.
(4) Bohm, D. J., Hiley, B. J. and Stuart, A. E. G. *International Journal of Theoretical Physics* (1970) **3**, 171.
(5) Einstein, A., Podolsky, B. and Rosen, N. *Phys. Rev.* (1935) **47**, 777.
(6) Bohr, N. *Atomic Theory and the Description of Nature* (Cambridge University Press, 1934), 92.

(7) Aharonov, Y. Private communication.

(8) Hilton, P. J. and Wylie, S. *Homology Theory* (Cambridge University Press, 1960).

(9) Misner, C. W. and Pheeler, J. A. *Annln. Phys.* (1957) **2**, 525.

(10) Frank, F. C. *Proc. Phys. Soc.* (1949) A **62**, 131.

(11) Bohm, D. J. *Satyendraneth Bose. 70th Birthday Comm.* II (1964), 280.

COHOMOLOGY OF OBSERVATIONS

R. H. ATKIN

Introduction. I wish to claim that the property of the space-time structure that really matters is its *connectivity* (which is a physical-observation thing) and that this connectivity is not necessarily the topological one (in the analytical sense) of the mathematician's R^4. On the contrary we should look for an expression of the essential connectivity via the concepts of *algebraic* topology, that is to say, in the language of homology/cohomology. When we do this we notice that the classically-assumed space-time structure is homologically trivial and that we can reproduce an homologically equivalent space-time structure without having to assume the continuum of the reals. Thus, homologically speaking, a discrete set of observations can be equivalent to a continuum. This is achieved mathematically by the use of Čech homology on a set, the definition of which requires (only) 'a covering' and not 'a topology'.

If such a structure is accepted as an expression of physical observations of space-time then the dual theory of cohomology expresses the 'other' possible observations and the homological structure of space-time (the so-called 'backcloth') determines the 'natural laws' of physics in general. This will require a replacement for Lorentz covariance—which in fact (in its day) amounted to replacing a topology (Newtonian) by a covering (Einsteinian). This replacement amounts to the algebraic invariance of the homological structure of space-time and, in particular, the one observed by the light signal.

1. An analysis of our traditional thinking. When we come to think about the problems of scientific observation (by which I mean the generally accepted body of 'experimental facts' together with the equally accepted methods of acquiring them) it is extremely difficult to free ourselves from a long tradition which has trapped us into making an extraordinary number of assumptions. This tradition certainly goes back to the scientific thinking of the Renaissance,[1] and because of this it would appear to stretch back even to the ancient Greeks.

Of course, the precise nature of this tradition is to be found in an analysis of the assumptions which have guided it and which have hardly been questioned until the early twentieth century (although one should not overlook the profound questioning which lay at the root of Hamilton's discovery of the algebra of quaternions, as he searched for a language with which to discuss natural philosophy).

These assumptions form a complex pattern and range over questions which on the one hand are essentially philosophical whilst on the other they appear to be primarily methodological. It is this latter aspect which is the subject of this paper, and we shall begin by considering the assumptions which lie at the heart of what is normally referred to as Newtonian physics.

(i) In the first instance the Newtonian claim must be that 'physics' is concerned with a certain *class* of 'observations' which we (the observers) attribute to an objective world. An observation is very much tied up with the 'act of observing'—a process which starts with the careful exercise of our unaided senses and proceeds to situations which ultimately include pieces of apparatus and scientific instruments. Some analysis of observation in terms of a 1–1 correlation between phenomenal causation and sensory perception is to be found in the literature.[3], [4]

This class of observations contains (mathematical) *sets* of observations to which we ascribe specific nouns, and we say that these nouns are the names of things or qualities in the objective world. In this way we form a correlation between our intuitive ideas about the world and the results of (scientifically) observing it, and we try to iron out the ambiguities of language which normally express the former by using a different language (viz. mathematics) to express the latter. This mathematical expression of the results of scientific observation then gives the subsequent theorising an intellectual respectability, which is desirable; sets the scientific studies on a logical basis, which is not true; and results in our science giving us an objective, impersonal, rational and reliable account of the phenomenal world; and thus is naive. The problem here is essentially one of choosing the language and whilst selecting mathematics (Newtonian-type mathematics) admirably provides precision and logical consistency it is easy to overlook the elaborate mathematical structure (the grammar of the language) which is also assumed.

But this is a later problem which we shall discuss subsequently; for the moment we examine this idea of a set of observations being at the root of a physics.

Suppose, for example, that there is a class of observations which correspond to our intuitive idea of 'geometrical position'—then we are saying that a physics of geometrical position exists when we insist that these observations form a mathematical set. That is to say, about any observation we must be able to say that we know whether or not it is such a one, viz., it indicates this idea or not, and between any two members of this set of observations there is a clear distinction— the possibility of *discrimination*. How the set is obtained in the first place, through what subtle use of imagination or apparatus, is proper matter for experimental research and is *independent of this assumption of the set-property*.

This set-property means that it is not permissible to speak of (say) 'mass' as something *defined* by all its possible measurements: the concept of mass does not come *after* its observation, but is an essential part of that observation and corresponds to set membership. The set-property is an expression of the fact that there is a profound sense in which we must know what we are observing as we observe it; in its turn the set-property is manifest by the *practical agreement* between experimental practitioners as to what they are about. The 'cleaning of the teeth' is not in fact to be confused with 'weighing a pound of sugar'.

In an earlier paper [2] I have used the word *scale* for any set of circumstances which enable an observer to collect a specific set of observations, and in this light we may regard much of (experimental) scientific research as the pursuit of scales of varying subtlety. The word *scale* therefore includes both naive and elaborate experimental arrangements, subject only to the constraint that the observer (who is part of the scale) knows what is 'observable' thereon (the set-property).

What in fact is observed 'on' (or via) a scale I have previously called an *element* of the physics, and since it is closely tied up with the scale we shall denote the latter by $S(E)$; E denoting the element.

We must now say that 'a physics' will be a study defined by a collection of elements and sets of observations on associated scales. Thus we should contemplate a *physics of space*, a *physics of particle dynamics*, a *physics of electromagnetic phenomena*, etc. (see reference (2), where some discussion of nine classical examples of a physics is given).

Each such physics will be adequately defined in terms of a *basic finite set of elements* E_i ($i = 1, 2, \ldots, k$). On this basic set will be built other concepts which will be subject to 'observation'—but essentially in terms of the original k elements. Such a basic set will be called a *set of generating elements* for the physics. Thus, in particle dynamics, if length and time are regarded as generating elements then 'velocity' is an element with observational content, but is not a generating element.

Furthermore, among the k elements of the basic generators, we must distinguish a subset (which is to contain at least one member) which shall be called *base elements*—designated by E_0^i ($i = 1, \ldots p \leqslant k$). These naturally arise once we notice that to assert the observation of an E on the scale $S(E)$ is to assume more about the S than we would like to do, viz., that such an observation involves postulating at least one element which is an *essential part* of the scale S. Recognition of a base element is equivalent to recognition of a scale: thus we cannot observe 'position' via marks on a ruler without the aid of a seeing-agent, or signal, which is normally called 'light'; this element would be a base element for this scale.

At this stage in the analysis it is clear that we are not speaking of 'measuring' elements in the accepted manner—which involves attaching numbers or other mathematical symbols to them—but our references to 'observations' are pre-mathematical. It is important to know at what precise point we make a move into the mathematical language; historically the move took many centuries of thought and was even then left open to criticism only because of the paucity of mathematical choice available; it is ironic that our students now make the move almost absent-mindedly.

Naturally, we are most likely to interpret as base elements the sense-stimuli which are involved in scales, but the concept here introduced is not to be narrowly restricted to this class. Thus it seems unavoidable to demand that the physics of electromagnetism is observed on scales for which 'electric charge' is to be a base element: chiefly its function is to act as 'signal' for manifestations of itself, and yet these manifestations are not adequately defined in terms of the five senses.

It now follows that we should denote the scales of a physics by $S(E_0^i; E_j)$ where the E_0^i are the base elements for this scale.

The various kinds of physics which will arise will now depend on the choice of scales, the choice of base elements, and the choice of generating elements. In particular, what is a base element for one set of scales need not be a base element for another. The idea of a physics (pre-mathematical) may then be expressed in:

DEFINITION: Given a set of elements $P_k = [E_1, E_2, \ldots, E_k]$ and a set of base elements $P_0 = [E_0^1, \ldots, E_0^h]$ associated with scales $S_\alpha(P_0; P_k)$, then the set of observations on the scales defines a physics of the P_k over the $S_\alpha(P_0)$.

It is an important feature of Newtonian physics that *base element* is to play no part; the first important gesture away from this position came when 'light' as a signal was singled out in Einstein's Special Theory of Relativity.[16]

(ii) It is commonly said that Newtonian physics presupposes a certain simple kind of space-time; then for example, particle dynamics is the study of a 'particle' which moves in this space relative to this time. Observation of its velocity, acceleration and forces-acting-on-it complete the physics by giving us a mathematical formulation of the interdependence of these elements.

Certainly Newtonian physics begins with a certain *idea of space*, but it is certainly not a *physics of space*—as we have defined above.

The idea of space was that the classical Greek geometry, with its Pythagorean metric and its consequent real number continuum, **R**, *is* space—that is to say, is identical with the physics of space which we would obtain if we made the effort to do so.

Unfortunately for this point of view it is not possible to verify the Pythagorean metric on any relevant scales since the irrational numbers elude observation (as the ancient Greeks well new). It followed that observations of certain lengths in the assumed physics of space would demonstrate only an approximate verification. Newtonian physics, in its initial assumptions, therefore acquired the characteristic that observations were inevitably approximate; this in turn has strengthened the feeling—which is presumably Platonic in its origin—that there is an *ideal*, or *absolute*, world which scientists must always find elusive.

This absolute world of observations finds its expression, as far as Newtonian physics is concerned, in the assumption of the significance of the real number continuum **R**. This significance occurs in two senses, viz., an *algebraic* sense and a *topological* sense,[17] and the Newtonian tradition can be expressed by saying that the physics of space is homeomorphic to **R**³ with a metric topology $\mathcal{M}$ under a bijective map

$$\sigma : \mathcal{S} \to (\mathbf{R}^3, \mathcal{M})$$

where $\mathcal{S}$ denotes the observations of space geometry. This map is also to be associated with another bijection α_σ which ensures that $\mathcal{S}$ has the algebraic structure of $\oplus^3\mathbf{R}$, viz.,

$$\alpha_\sigma : \mathcal{S} \to \oplus^3\mathbf{R}.$$

This same pattern can be seen in the Newtonian attitude to the physics of time—whose essence, observation-wise, is an appeal to our intuitive sense of a *total ordering*. In this case the bijective map

$$\tau : \mathscr{T} \to (\mathbf{R}, \mathscr{U}),$$

where $\mathscr{U}$ denotes the usual topology on $\mathbf{R}$, is associated with the 'algebraic' map α_τ where

$$\alpha_\tau : \mathscr{T} \to (\mathbf{R}, <).$$

Newtonian space-time is now characterised by the topology of the product space $\mathscr{S} \times \mathscr{T}$; this topology is the *product topology*. It is equivalent to saying that there is no intrinsic relation between space and time observations; there is no signal, no base element in the observations.

We shall refer to this space-time with its product topology as the *Newtonian backcloth*; against this, as a background, physics developed in various directions,[2] but the generating elements for the backcloth are to be found in every other such physics.

The work of Hamilton was the first attempt to question the wisdom of using bijective maps into the *algebra* of the reals, but no attempt was made to question the topological map σ until the beginnings of this century and the work of Einstein and others in Relativity Theory.

(iii) We can now see that attempts to modify Newtonian physics will be germane in the following contexts.

(a) Abandon the 'absolutism' theory; permit the scales $S_\alpha(P_0; P_k)$ to define the physics (cf. reference (12))—particularly in the case of the space-time physics.

(b) Accept the space-time physics as a backcloth but abandon the classical assumption about the reals: adjust the product topology so as to take cognisance of the base element(s).

(c) Abandon the space-time physics as a backcloth; replace it by another.

(d) Accept the Newtonian backcloth but abandon the *algebraic* dependence on the reals.

The possibility of pursuing (c) is hardly likely to appeal to practising physicists, but it is interesting to note that other disciplines (e.g. economics, sociology) which try to base their theories on observations might well feel freer to experiment with different backcloths.

On the other hand one would expect the experimentalist to rise to the challenge of (*a*), taking an opportunity to produce a physics of space—in the sense of our definition of a physics—and thus providing a sounder methodological base for classical physics.

Under (*b*) we can clearly include the Einstein–Lorentz theories, based as they are on the introduction of *pseudo-metrics* on the space-time observations.

The pseudo-metric $ds^2 = c^2 dt^2 - dx^2 - dy^2 - dz^2$ of special theory could only be justified by stressing the role of the base element (which was 'light' for all the scales) and thereby undermining the absolutism of Newtonian physics. But the reals were not entirely rejected, since the $x, y, z, t \in \mathbf{R}$, so that the modification became an adjustment to the *topology* of the Newtonian backcloth: naturally we shall refer to the new backcloth as Einsteinian.

The pseudo-metric does not immediately induce a topology on the product space $\mathscr{S} \times \mathscr{T}$, but denoting the latter by $\mathscr{B}$, we may define an equivalence relation $\sim$ on $\mathscr{B}$ by setting $p \sim q, p, q \in \mathscr{B}$, whenever $m(p, q) = 0$, m denoting the pseudo-metric. We now have a metric induced on the quotient space $\mathscr{B}/\sim$, in which there is an identification of points in the same equivalence class. In the Einsteinian backcloth this identification is between points on the light cone, given by

$$c^2 t^2 - x^2 - y^2 - z^2 = 0.$$

Furthermore the Newtonian maps σ, τ will now be replaced by ϵ and α_ϵ where

$$\epsilon : \mathscr{S} \times \mathscr{T} \to (\mathbf{R}^4/\sim, \mathscr{M})$$

and $\mathscr{M}$ is now the metric topology induced by the Lorentz pseudo-metric on the quotient space $\mathbf{R}^4/\sim$. The associated algebraic map α_ϵ will be

$$\alpha_\epsilon : \mathscr{S} \times \mathscr{T} \to \mathbf{R}^4_A/\mathscr{I}$$

where $\mathbf{R}^4_A$ denotes the algebra of the reals over the 4-space and $\mathscr{I}$ denotes the *algebraic ideal* generated by the quadratic form $c^2 t^2 - x^2 - y^2 - z^2$, in that algebra.

This means that the introduction of the Lorentz–Einstein pseudo-metric induces an algebraic representation of the physics of space-time which is a *Clifford algebra* (associated with the quadratic form). [18] It is clear that the same situation applies to General Theory; it is interesting to note that an entirely algebraic discussion of the main results of General Relativity was provided as long ago as 1921. [8]

We notice also that the Clifford algebra associated with Special Relativity contains, as sub-algebras, the quaternions $\mathscr{Q}$, the complex numbers C, and the reals R. A generalisation of the usual formulation of Special Theory (which uses R) to one involving C is therefore compatible with this analysis, and in a previous paper [5] I have used this to distinguish the case when c, the velocity attributed to the base element, is not itself observable by that base element.

(iv) With the beginnings of Quantum Theory came the acknowledgement of the fact that observations on a scale $S(E_0; E)$ are essentially *discrete* and *countable*. The role of Planck's constant h was first and foremost to replace the reals R, as an image space for observables, by a countable subset, viz. $[nh; n \in Z$, the integers$]$.

But we notice that no such assumption was made about the space-time backcloth $\mathscr{S} \times \mathscr{T}$. In fact both the Newtonian and Einsteinian backcloths have been accepted in quantum theory, giving rise to non-relativistic and relativistic theories. Considering the emphasis placed on observable-based theories by Heisenberg and others it is surprising that the backcloth was not scrutinised more critically.

In the non-relativistic quantum theory the Newtonian backcloth is accepted as a topological representation of $\mathscr{S} \times \mathscr{T}$, whilst the other elements (momentum, energy) are 'quantised'. This means that they are to be represented by functions in the class $Z^{\oplus 4R}$, Z being identified as the integers up to isomorphism.

In relativistic quantum theory an attempt is made to use the Einsteinian backcloth together with the Clifford algebra $\mathscr{C}_4$. The energy and momentum elements are now represented by functions in the class $Z^{\mathscr{C}_4}$.[19]

In either case the main characteristics of the theory are to be found in justifications for the restrictions of functions in $R^{\mathscr{B}}$ to $Z^{\mathscr{B}}$. For example, classical variable represented by a map $f \in R^{\oplus 4R} \Rightarrow f \mid Z^{\oplus 4R}$ represents a quantised variable.

An important feature of the Quantum Theory is the fact that (e.g.) *momentum* has become a new *generating element* in the physics. In Newtonian and Einsteinian physics momentum (and energy) is (only) an element and is defined by observations on the appropriate space-time backcloth. The translation to Quantum Mechanics is effected by the use of the classical definition as a weaker 'classical analogue'— a move which is intuitively difficult to accept and which disguises the new *generating* role of the element.

In classical mechanics momentum is only defined in terms of the generating element 'mass' (and of course $\mathscr{S} \times \mathscr{T}$) but in Quantum Mechanics this is not so.[11]

The mathematical patterns in the two cases are now:

Non-relativistic QM: Newtonian backcloth for space-time; representation of $\mathscr{S} \times \mathscr{T}$ by functions in $\mathbf{R}^{\oplus 4\mathbf{R}}$

Momentum as generating element; representation by functions in $Z^{\oplus 4\mathbf{R}}$

Energy defined via classical (Newtonian) analogue.

Relativistic QM: Einsteinian backcloth for space-time; representation of $\mathscr{S} \times \mathscr{T}$ by functions in $\mathbf{R}^{\mathscr{C}_4}$

Momentum as generating element; representation by functions in $Z^{\mathscr{C}_4}$

Energy defined via relativistic analogue.

The mathematical formulation of Quantum Mechanics in terms of self-adjoint operators on an Hilbert Space constitutes a non-trivial linear method of defining the restriction $\mathbf{R}^{\oplus 4\mathbf{R}} \to Z^{\oplus 4\mathbf{R}}$. It is at this stage that the mathematical algorithm first enunciated by von Neumann (see also reference (11)) may be introduced.

Some early divergences in Quantum Theory—e.g. the infinite zero-point energy of radiation in a rectangular cavity, or the transverse self-energy of an electron—certainly led to attempts at 'quantising length' as a possible way out. This indicates some attempt at the reform listed as (a) above; a valuable discussion is to be found in reference (13).

2. The homology of observations. The analysis in the first part of this paper has shown that topological structure in sets of observations, whether deliberately or accidentally injected into the theory, has played a significant role in physics. But topological structure, in the sense of a set possessing 'a topology', is quite a complicated structure, and so it would be desirable to look for a more fundamental kind of structure associated with sets of observations—and one which transcends any associated algebraic formulation of the topology, when that exists.

I wish to suggest that such a fundamental view of sets of observations can be obtained via the notion of *homology*,[6] and that the development of physics to-date is based on a far-reaching *assumption* about the homology of the space-time backcloth.

In the case of classical physics, more fundamental than any mention of metric is the assumption that space-time $\mathscr{S} \times \mathscr{T}$ is *homologically trivial* (*acyclic*), that is to say, given any covering $\mathscr{A}$

of the postulated space-time, the Čech homology groups [6], [9] are trivial,

$$H_p(\mathscr{S} \times \mathscr{T} ; \mathscr{A}) \cong 0, \quad p > 0.$$

Loosely speaking, there are no holes in $\mathscr{S} \times \mathscr{T}$.

The most important feature of this property lies in the fact that it may still be true (that is to say, it can have just as much meaning either as an assumption or as an observational fact) whether or not the space-time $\mathscr{S} \times \mathscr{T}$ is embedded in the classical Euclidean $\mathbf{R}^4$ or not. In other words, homological structure is pre-topological structure; essentially, the idea of a *covering* of a set is independent of the idea of a *topology* on a set.

If we consider a set of observations which we shall agree to call space-time (and let us use the word 'point' without prejudice) then we may obtain a representation of the Cartesian product $\mathscr{S} \times \mathscr{T}$ by a diagram as shown (fig. 1).

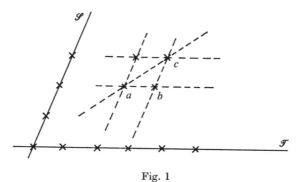

Fig. 1

The points which are observed as $\mathscr{T}$ and those which are observed as $\mathscr{S}$ produce a mesh on $\mathscr{S} \times \mathscr{T}$ as indicated.

Now the absolutism of the Newtonian physics regarded these points as Euclidean points and the whole diagram embedded in the Euclidean plane $\mathbf{R}^2$ and it accepted that plane as the 'objective' or 'real' space-time. This view would interpret the mesh as a possible triangulation (or cell decomposition) of $\mathbf{R}^2$; it would produce a *simplicial* decomposition in which the 0-simplices would be identified with the points of the mesh, the 1-simplices by the pairs of points of the mesh, generally the p-simplices would be defined by collections of $(p+1)$ points of the mesh.

But such a simplicial view of the diagram is not dependent on there being any other points in the plane than the ones marked. In speaking of a triangulation of $\mathbf{R}^2$ the 1-simplex defined by $[a, b]$ would be the

Euclidean line which joins the points a and b; but if we take away this underlying Euclidean support, and no longer speak of a triangulation of $\mathbf{R}^2$, then the subsequent homological analysis is exactly applicable to the notion that, for example, the 1-simplex $[a, b]$ is simply the set with two members, viz. a and b.

It follows that the assumption of the acyclic nature of $\mathscr{S} \times \mathscr{T}$ does not depend on the embedding in $\mathbf{R}^2$, nor indeed on whether the observation points are infinite or finite. Certainly the idea of a metric topology on $\mathscr{S} \times \mathscr{T}$ is not required prior to a discussion of the homological structure; although the Newtonian (and for that matter the Einsteinian) backcloth ensures the acyclic property. It is relevant here to observe that our unaided senses appear to give us a non-metrical view of space. [10]

Let us now suppose that the mesh diagram represents an observational backcloth—which is a genuine physics. Then the extension of this backcloth to a physics of dynamics will be achieved by (say) identifying a 'particle' with its 'point' in the mesh—and this in no way presupposes any idealistic view of what either 'point' or 'particle' means. The additional elements of the dynamical physics, such as 'mass', 'velocity', 'acceleration', etc., will now have a theoretical expression against this backcloth.

If we move along lines of thought which are parallel to those of classical physics then we would wish to associate 'mass' with the 0-simplices of the *complex* $K(\mathscr{S} \times \mathscr{T})$; that is to say, when the particle is observed to be coincident with a point in the mesh diagram then it is observed to have this thing called mass—at that point. Attributing 'values' to this mass (or giving it 'co-ordinates' on an abstract axis) amounts to introducing an algebraic structure, say, a *ring R*, and then introducing a map

$$m : C_0 \to R$$

where C_0 is the usual zero-order chain group on the complex K (that is, a free group generated by all the 0-simplices of K).

This mapping 'm' is the theoretical expression of the mass of the thing called a particle; its values are to be found in the algebraic ring R.

In the usual language it is an example of a *cochain* (dual to the idea of chain) defined on the complex K with R acting as coefficient group.† It is more usual to denote this by a symbol $c^0(K; R)$.

In the case of Newtonian physics the cochain map c^0 would be

† This formulation is too crude; further analysis shows that the concept of a coefficient *sheaf* is more appropriate (see reference (9)).

a constant map; the particle is supposed to have a constant mass at all points of the space-time backcloth.

Similarly, in parallel with the classical method we would expect to describe the element 'velocity' of this particle by associating a value in R with a 1-simplex (for example, with the pair of points a and b in the mesh diagram); this is because velocity refers to change of position relative to time. Thus 'velocity' must be regarded as a *cochain* of order 1, a map $c^1(K; R): C_1 \to R$.

The classical kind of linear momentum is a new element which is in some way associated with a product of the mass and the velocity. Since this momentum is another example of a 1-cochain, a c^1, we may most easily identify it as a map from the tensor product of cochain groups C^0 and C^1 viz., $\quad \text{mom}: C^0 \otimes C^1 \to C^1.$

Generally, we would now expect to be able to represent any element in our physics by the theoretical p-cochain $c^p(K; R)$, where p is a non-negative integer. The case of $p = 2$ occurs with the classical concept of angular velocity or angular momentum; $p = 3$ occurs with (oriented) volume, density, etc.

The case of Newtonian or Einsteinian physics corresponds to elements associated with maps $\{c^p\}$ with values in the reals $\mathbf{R}$. Quantum Mechanics, on the other hand, is clearly taking Z (the integers) as the ring of coefficients.

This discussion also brings out very clearly the point which is made by Quantum Theory (but we now see that it is not peculiar to that theory) viz., that classical variables correspond to linear operators (on a suitably linear space). These operators are the cochain maps on a suitable (linear) chain complex K and into an appropriate coefficient group, in the classical case; and they are maps into the complex K itself, in the quantum case. Further analysis of this gives us more insight into the role of the so-called wave functions in the typical Hilbert space theory; details will be given in a later paper.

If we return to the mesh diagram and consider again the mass and velocity cochains, we see that it is important to keep the order of the cochain clearly in mind when we are comparing the two. In particular, the classical idea of linear momentum as the product (mass × velocity) necessitates that mass should be a constant map when velocity is a c^1—otherwise the momentum is not well-defined as a c^1. On the other hand, if we try to avoid this situation and permit ourselves to have a mass (c^0) which is not constant at all vertices, then the momentum (c^1) is only well-defined if we can somehow attribute velocity to a *vertex* only. This means that velocity must become a c^0 and so must the momentum. This can only be achieved by altering the mesh diagram so that all neighbouring vertices are allowed to coincide: such a process is the traditional limiting exercise—which we see reduces the mesh plane to the Euclidean plane (and requires the coefficient group to be complete). The price we would have to pay to achieve this would be the burden of inventing the differential calculus.

This suggests that the successful development of Newtonian physics and the parallel triumph of the mathematical tool—the differential calculus—were not merely coincidental.

The difficulty of matching up the mass and velocity cochains points to the same quandary with the *generating* element momentum (c^1) in quantum mechanics and its counterpart position. This is expressed by the Uncertainty Principle: given the momentum defined on the simplex $[a, c]$, what is the position to be associated with it—a or c?: given the position $[b]$ in the complex, what is the momentum to be associated with it—$[a, b]$, $[bc]$, etc.?

Now if we return to the fundamental assumption of the acyclic homology on the backcloth (and we have seen that this is not dependent on the assumption of either Euclidean geometry or of the reals **R**) we can see how a *natural law* is automatically built into the physics.

Assuming $H_p(K) = 0, p > 0$, then if c^p is an observable element in the physics, z_{p+1} any cycle in the backcloth, ∂ and δ the usual adjoint boundary and coboundary operators,

$$0 = (\partial z_{p+1}, c^p) = (z_{p+1}, \delta c^p), \quad \text{all } p > 0.$$

Hence we would observe all our c^p in such a way that, on every cycle (cf. closed curve or surface)

$$\delta c^p = 0 \tag{1}$$

which is the inevitable natural law for acyclic backcloths.

If the cohomology is also trivial, there would be, for every element c^p another observable element (its *potential*) c^{p-1} such that

$$c^p = \delta c^{p-1} \tag{2}$$

These equations (1) and (2) summarise all we can expect from this kind of backcloth. They are amply illustrated in theoretical physics, albeit somewhat hidden under various formulations of field theories, and they arise in the following way via de Rham theory of differential forms.

We set up a 4-dimensional vector space V_4 isomorphic to the classical manifold $\oplus^4 \mathbf{R}$; in fact a basis for V_4 is provided by the differentials dx, dy, dz, dt. The *exterior product space* $\wedge V_4$ is then constructed and formal cochains are identified as differential forms (of various orders p) associated with $\wedge V_4$. The *exterior derivative* acts as the coboundary operator on these cochains, the corresponding cohomology groups being denoted by D^p.

Now it is known [14] that if X is a compact space and A a cover of X, then the de Rham cohomology is isomorphic to the Čech cohomology which is based on the complex $K(A)$ of X, that is

$$D^p(X) \cong H^p(K; \mathbf{R}).$$

This means that, although the field equations (which we shall see are contained in the de Rham theory) have been built up out of the limiting ideas of differentials on a continuum, they are in 1–1 correspondence with and have the same algebraic structure as Čech cochains which satisfy (1), and sometimes (2). These cochains may be thought of as being defined against a backcloth which is not necessarily continuous—since the topological structure usually associated with the coefficient group $\mathbf{R}$ is irrelevant.

Examples of equation (1) include the usual *equations of wave motion*, the *equations of diffusion*, the *equations of heat conduction*, the *equations of fluid mechanics*, *Maxwell field equations*, *Hamilton's equations* for a dynamical system, etc. (see references (14), (15)).

As an illustration, Hamilton's field equations are derivable from a differential form

$$c^1 = p_1 dq^1 + p_2 dq^2 + \ldots + p_n dq - H\, dt$$

where the q_i, p_i are the usual canonical variables and H is the Hamiltonian function. The exterior derivative may be defined as a linear map $\delta: C^p(\wedge V_n) \to C^{p+1}(\wedge V_n)$, where $\wedge V_n = \bigoplus_q \wedge^q V_n$, satisfying

   (i) $\delta(c^p \wedge c^q) = (\delta c^p) \wedge c^q + (-1)^p c^p \wedge (\delta c^q)$

   (ii) $\delta c^0 \equiv dc^0$

   (iii) $d^2 c^0 \equiv 0$

$\wedge$ denoting the exterior (antisymmetric) product. The Hamilton's equations are equivalent to the equation (1), using the cochain defined above. The significance of a *contact transformation* amounts to changing the cochain c^1 into any other to which it is co-homologous.

If we look for the existence of a 'potential', so that equation (2) is satisfied, then we seek a 0-cochain c^0 such that the given $c^1 = \delta c^0$.

This means that we seek a function $W(q^i, t)$ such that $c^1 = dW$. Hence we shall require $p_i = \partial W/\partial q^i$ for $i = 1, \ldots, n$, as well as

$$\partial W/\partial t + H = 0$$

which is the well-known *Hamilton–Jacobi equation*.

3. Analysis of scale. We have seen that a scale $S(P_0; P_k)$ will either provide us with a *backcloth* of observations or with a ring of cochains which are themselves defined on a backcloth. But the difference is not as profound as it might seem, since the members of the chain groups $C^p(K)$ on a simplicial complex K may themselves be interpreted as cochains.

The decision as to whether or not a scale $S(P_0; P_k)$ shall be regarded as a backcloth is therefore one which can be settled pragmatically.

The following definitions and axioms are offered as a preliminary process to setting up a convenient backcloth, that is to say, to describing it in a formal mathematical language. In the first place we must establish the conditions for describing a *covering* of a set of observations of some element E (this is the pre-topological situation). Secondly, we must consider what is involved in introducing some *algebraic* structure into the set E.

DEFINITION 1: The members of the set of phenomena $S(E_0; E)$ are the *existence states* of E on $S(E_0)$.

If we denote this set by $\bar{E}$, and if $\bar{E}$ is indexed by a set I, so that $\bar{E} = \{E_i; i \in I\}$ then we require that E_i and E_j should correspond to distinguishable states whenever $i \neq j$.

This process is the fundamental one of *discrimination* between observations which is referred to by E. W. Bastin in an accompanying paper.

We now distinguish between the *idea* of the existence states (which occurs at the intuitive level of observation) and the practical act of observation (which is the beginning of scientific observation) by:

DEFINITION 2: An *observable state* on $S(E_0; E)$ is a subset $X \subset \bar{E}$.

AXIOM I: Each existence state E_i is manifest as an observable state via the singleton set $\{E_i\}$.

AXIOM II: If Ω denotes the class of all observable states on $S(E_0; E)$, then $X \in \Omega$ and $Y \subset X$ imply $Y \in \Omega$.

It follows that the empty set sign $\varnothing$ is an observable state on every scale (being the absence of observation), and Axiom II also implies that '$X, Y \in \Omega$ implies $X \cap Y \in \Omega$'—observability is closed under Boolean intersection. But notice that we do not demand that *all* subsets of $\bar{E}$ should be in Ω; in particular $\bar{E}$ may not be an observable state.

These axioms are sufficient to define a covering of the set of observations, which in turn defines a simplicial complex $K(\bar{E})$.

We notice that a structure very similar to the one proposed was introduced by A. N. Whitehead in 1920.[12]

It now follows that Ω may be used to generate a *topology* $T(\Omega)$ on the set $\bar{E}$. Since each 'point' E_i of $\bar{E}$ is open in T it follows that T is the *discrete topology* on $\bar{E}$.

In the particular case of there being only one existence state E then the observable states are $\{\phi, \bar{E}\}$ which is the indiscrete topology $\mathscr{I}$ on $\bar{E}$.

DEFINITION 3: A scale $S(E_0; E)$ with the indiscrete topology is a *primitive scale*, whilst a scale for which $\Omega \equiv T(\Omega)$ is a *complete scale*. We shall now consider two axioms which we shall bear in mind as alternatives; they refer to countability.

AXIOM IIIA: For any scale $S(E_0; E)$ the set of existence states $\bar{E}$ has finite cardinality.

This implies that there is always an index set I which is a finite subset of the positive integers Z^+. If card$(\bar{E})$ denotes the cardinality of $\bar{E}$ then there exists a positive integer n such that card$(\bar{E}) = n$.

AXIOM IIIB: For any scale $S(E_0; E)$ the set $\bar{E}$ is infinitely countable, that is to say, card$(\bar{E}) =$ card(Z).

We have noticed that the set Ω defines a covering of the set $\bar{E}$ (and if we wish to make an oblique reference to the fact that Ω is a subset of $T(\Omega)$ we could call Ω an *open* cover). Hence we have all the machinery available for constructing a simplicial complex $K(\bar{E})$ and for setting up a typical Čech homology $H_p\{K(\bar{E}); Z\}$ to be associated with E.

We also notice that this homology will be trivial (*acyclic*) when S is a complete scale.

In any event we may now consider the possible cochains on this backcloth as defining suitable observable elements—which might arise in setting up a physics other than that of the (pure) backcloth. Since this concept is to be basic to our analysis we assume:

AXIOM IV: Given a scale $S(E_0; E)$ and an element E', then there exists a *sheaf* of rings $\mathscr{R}$ such that, relative to the backcloth $S(E_0; E)$, the observable states E' are represented by the cochains

$$C^p\{K(\bar{E}); \mathscr{R}\}.$$

Here we see that a certain algebraic structure is presumed to be available and to be relevant to the observations of E', but the Axiom indicates the *relative* role of the algebraic properties whilst giving a *fundamental* role to the homological structure of the backcloth.

It is noteworthy that this subordination of the algebraic to the homological (and therefore to the topological) was a characteristic of the thinking of both Eddington and Whitehead. In Eddington's case it gives us some understanding as to why his thesis (which said roughly, that physics is 'all in the mind') aroused such a mixture of attraction and repulsion in the scientific public. The freedom which the phrase conjures up is, from our point of view, the freedom of the algebraic choice (if there is one), but after this choice has been made there comes the topological constraints inherent in the homological structure—that is where 'reality' lies. The work of Whitehead was more obviously concerned with a basis (in a technical sense) of a topological structure on a fixed backcloth (that of space-time); he did not experiment as boldly with algebraic structure as did Eddington.

Let $\mathscr{A}$ denote an algebraic structure (group, ring, field, algebra) and consider the problem of searching for a 1-1 mapping

$$\sigma: \mathscr{A} \to \Omega'.$$

When we have found such a mapping we shall say that the structure $\mathscr{A}$ has been imposed on the set Ω'.

PROPOSITION I: A semi-group $\{G; \oplus\}$ may be imposed on

$$\{\bar{E}', \Omega'\} \text{ iff card} (\Omega') \geqslant \text{card} (G).$$

It is clear that the condition is both necessary and sufficient for the existence of a 1-1 map $i: G \to \Omega'$, and that i is not unique when card $(\Omega') > 1$.

This mapping i induces a binary relation $\oplus_{\Omega'}$ on Ω' via the commutativity of the accompanying diagram (fig. 2).

DEFINITION 4: When such an injection i exists we say that $S' \equiv \{\bar{E}', \Omega'\}$ *carries* G.

COROLLARY: S' carries a group (G), ring (R), algebra (A) iff card $(\Omega') \geqslant$ card (G), card (R), card (A).

We see that S' becomes associated with a class of algebraic structures, viz., those which it carries.

We notice also that Ω' is partially ordered, by inclusion, whereas the set $\bar{E}'$ (viewed as a set of *singletons* $\{E'_i\}$) does not possess any such inherent property. It follows that we obtain the greatest freedom structure-wise if we can inject an algebraic structure into $\bar{E}'$. This will be possible if card $(\bar{E}') \geqslant$ card (A).

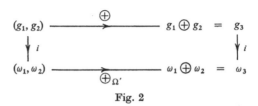

Fig. 2

DEFINITION 5: We shall say that the scale $S' \equiv \{\bar{E}', \Omega'\}$ *supports* A when card $(\bar{E}') \geqslant$ card (A).

We notice that although *supports* implies *carries* it is not true that *carries* implies *supports*.

PROPOSITION II: Every primitive scale carries a Boolean algebra isomorphic to Z_2 (the ring of integers (mod 2)).

This follows by the map $i: \{0, 1\} \to \{\phi, E'\}$.

PROPOSITION III: Every complete scale carries a Boolean algebra isomorphic to all Z_2-valued functions.

This follows by mapping the members of Ω' into the characteristic functions of the subsets of $\bar{E}'$.

PROPOSITION IV: Every complete scale possesses a decomposition into a set of primitive scales.

We need only define these primitive scales S'_i by the sets $\{E'_i, \mathcal{J}\}$ where $\mathcal{J} = \{\phi, E'_i\}$; then $S' = \cup S'_i$ and $S'_i \cap S'_j = \{\phi\}$ when $i \neq j$.

In physical terms the primitive scales are the circumstances under which we recognise the observable nature of the existence states.

PROPOSITION V: The Boolean algebra of Proposition III is isomorphic to $Z_2 \oplus Z_2 \oplus \ldots$ where the number of factors equals card $(\bar{E}')$.

COROLLARY: Under axiom IIIB we notice that the algebra of Proposition V is isomorphic to **R** (as binary arithmetic).

PROPOSITION VI: Under Axiom IIIB a scale S carries the reals $\mathbf{R}$ but does not support $\mathbf{R}$.

It is a legitimate exercise to consider the problem of increasing the cardinality of the scales at our disposal in order to accommodate some specific algebraic structure. Such a procedure, which is an experimental problem, amounts to a *refinement* of the covering Ω (and of T) in the usual mathematical sense.

There are two basic methods of achieving this:
    (i) a process we shall call magnification;
    (ii) a process we shall call resolution.

DEFINITION 6: We shall say that a scale S^1 is a *magnification* of the scale S when each primitive scale (or singleton) is identified with the union of k primitive scales (or singletons) on S^1.

This may also be described by saying that Ω^1 is decomposed into equivalence classes and then $\Omega \leftrightarrow \Omega^1/\sim$.

If we use the word particle to describe what is represented by a singleton set then magnification means that 'particle on S' is seen as 'k particles on S^1'. The process corresponds to the discovery of a certain kind of *fine structure* in the corresponding element.

DEFINITION 7: When all the observable states on a scale S become the singletons on a scale S^1 we shall say that S^1 is a *resolution* of S.

We see that resolution is not the same thing as magnification since:
    (i) under Axiom IIIB resolution is not possible, since it would be necessary for $\mathrm{card}\,(S^1) = c$, the power of the continuum,
    (ii) under Axiom IIIA resolution is certainly possible, but if $\mathrm{card}\,(\bar{E}) = n$ and (in the worst case) the scale is complete, so that $\mathrm{card}\,(\bar{E}) = 2^n - 1$, then this can only be magnification if there exists a positive integer k such that $kn = 2^n - 1$: this requires the congruence $2^n \equiv 1 \pmod{n}$, which is false when $n > 1$.

This process of resolution is another kind of fine structure discovery and it has this interesting property that for complete scales under Axiom IIIB (and that means an acyclic homology) resolution is impossible. We therefore conclude that *when resolution exists* Axiom IIIA *must apply to the scales*.

We see too that Bastin's paper contains a discussion of resolution of scales, and discusses a certain mathematical algorithm which results in a limitation of the possible resolutions. Because of this we consider the consequences of

AXIOM V: For a scale S resolution is limited.

By 'limited' we shall mean that successive resolutions, giving rise to sets $\bar{E}^0$, $\bar{E}^1$, ... gives an *ascending chain condition*, that is to say, there exists an integer r such that

$$\text{(i)} \quad \text{card}(\bar{E}^{r+1}) = 2^{\text{card}(\bar{E}r)} - 1$$

and   (ii) $\bar{E}^{r+1}$ cannot be realised on a scale S^{r+1},

i.e. the process of successfully discriminating the observable states on successive scales breaks down; subsequent resolution is not practicable.

Such a limitation on resolution (under Axiom III A) must be an essentially structural property of the scales, and in this sense it is a function of the *base element* E_0 as well as of E. We might indeed consider a certain set of scales $S_\alpha(E_0; E_1, E_2 ...)$ which exhibit the same resolution limit $h = h(E_0)$; this h can be 'small' or 'large' depending on the algebraic structure.

If, for example, we contemplate a scale $S(E_0; E)$ where E is an element expressible in terms of two generating elements E_1 and E_2 in such a way that, in **R**, we have

$$x(E) = x_1(E_1)x_2(E_2)$$

then $h(E)$ will induce limits of resolution $h_1(E_1)$ and $h_2(E_2)$ where

$$h \leqslant h_1 h_2.$$

For example, take the Planck's constant h as a limit of resolution and consider the relation between the limits

$$h(\text{action}); \quad h_1(\text{mass}); \quad h_2(\text{time}); \quad h_3(\text{distance})$$

viz. $$h \leqslant h_1 \times h_2^{-1} \times h_3^2.$$

If mass and time are functions of the same base element (only) so that $h_1 = h_2$ then we obtain a limit of resolution for observations of distance (a lower limit of measurement) viz.,

$$h_3 = 0.814 \times 10^{-13} \text{ cm.}$$

At the other extreme, let us take a limit of resolution as the relativity limit $c = $ velocity of light. This, together with the previous result gives a smallest limit of resolution for time measures as

$$h_t = 0.27 \times 10^{-23} \text{ s}$$

an interval observable in connection with certain strong interactions between elementary particles.

REFERENCES

(1) Clagett, M. *The Science of Mechanics in the Middle Ages* (University of Wisconsin Press, 1961).

(2) Atkin, R. H. Abstract Physics, *Nuovo Cimento* (1965) **38**, 496–517.

(3) Basri, S. A. *A Deductive Theory of Space and Time* (North Holland, 1966).

(4) Henkin, L., Suppes, P. and Tarski, A. *The Axiomatic Method* (North Holland, 1959).

(5) Atkin, R. H. A generalisation of special relativity theory (Unpublished, 1966).

(6) Hilton, P. J. and Wylie, S. *Homology Theory* (Cambridge University Press, 1967).

(7) Eddington, A. *Fundamental Theory* (Cambridge University Press, 1948).

(8) Forsyth, A. R. The concomitants of quadratic differential forms in four variables, *Proc. Roy. Soc. Edinb.* (1921).

(9) Bredon, G. E. *Sheaf Theory* (McGraw-Hill, 1967).

(10) Zeeman, E. C. *The Topology of the Brain and Visual Perception* Prentice Hall, 1961).

(11) Mackey, G. W. *Mathematical Foundations of Quantum Mechanics* (Benjamin Inc., 1963).

(12) Whitehead, A. N. *The Concept of Nature* (Cambridge University Press, 1920).

(13) March, A. *Quantum Mechanics of Particles and Wave Fields* (Wiley, 1951).

(14) Goldberg, S. I. *Curvature and Homology* (Academic Press, 1962).

(15) Flanders, H. *Differential Forms* (Academic Press, 1963).

(16) Adler, R., Bazin, M. and Schiffer, M. *Introduction to General Relativity* (McGraw Hill, 1965).

(17) Hu, S-T. *Elements of General Topology* (Holden-Day, 1964).

(18) Chevalley, C. The construction and use of some important algebras, *Math. Soc. Japan* (1965).

(19) Dirac, P. A. M. *Principles of Quantum Mechanics* (Oxford, 1947).

THE ORIGIN OF HALF-INTEGRAL SPIN IN A DISCRETE PHYSICAL SPACE

TED BASTIN

1. Introduction. The writers of several papers in this book—Atkin, Bohm, Penrose, von Weizsäcker—have made reference to a view of physical space in which space is defined in terms of a finite number of points and in which a rule is postulated for constructing new points. I shall call a physical space of this kind *constructive*, by analogy with the constructivist mathematics of Brouwer. [1]

I believe that a constructive theory of physical space is potentially capable of resolving the clash between the continuum aspect and the discrete aspect of quantum physics, because to introduce it is a sufficiently radical change to eliminate our reliance on classical concepts. These have to be separately defined in terms of the new approach.

The basis for the present paper is a constructive theory [2] in which a construction rule for new points generates a hierarchy. The points at a given level of this hierarchy replace the quantum-mechanical eigenvectors, and an order-preserving mapping between the levels corresponds to a physically significant event. In a short paper in Part VI of this book, I argue that in a constructive space a sufficient condition to be able to define an operationally realistic concept of particle trajectory is that the points in the space be ordered, and therefore for the purpose of the present paper I shall summarize only so much of the constructive theory as is necessary to define an ordered set of points.

I shall then deduce the existence of binary vectors having the formal structure of half-integral spin in relation to rotation. I shall also compare the conventional and the discrete–constructive theories in respect of the extent to which they constitute an explanation of the half-integral spin.

2. The construction algorithm. I consider a set S of abstract *events* upon which a basic operation *discrimination* is defined. When two entities in S, A and B, occur in an interactive relationship together, one of them, say A, can discriminate between the two cases:

(1) that B is identical with A;
(2) that B is not identical with A.

213

Further, the result of this discrimination operation is also an event, C, in the set, and $C \neq A, C \neq B$; the set has therefore a construction process built in.

It has been shown by Amson, Parker-Rhodes and the writer [2] that if we wish to represent the entities in this set by ordered sets of binary symbols, 0 and 1, then the discrimination operation has to be Boole's original operation of addition, sometimes called *symmetric difference*. In an arithmetic context (which at present we are not in) this is addition mod 2. This operation is represented by the table

	0	0
0	0	1
1	1	0

It was also shown, by introducing a concept of closure under the discrimination operation, that the discriminable entities in the set are the linearly independent sets of vectors, excluding the null vector, of which there are $2^n - 1$, where n is the cardinal of the set of entities generated. This set of sets of entities is made the elements of a new *level* in which discrimination can again take place. Proceeding as before we get $2^{(2n-1)} - 1$ of these discriminated entities at the level after, and so on.

The discrimination process requires that the vectors, or sets of binary units upon which the discrimination process acts, be ordered. This requirement is satisfied as an axiom in a hierarchy whose simplest level consists of the vectors 0, 1 themselves. Henceforward I shall only discuss this kind of hierarchy. Since discrimination operations can take place at different levels and since the existence of entities at the different levels depend upon those at previous levels, it follows that a *record* of what entities exist has to be kept, independently of the set of entities. When n entities exist, this record would require n binary units in the memory, and since they have to be ordered the number of units becomes n^2. For the case $n = 2$, which we are considering, we therefore have our independent record system increasing as the sequence

$$2, 4, 16, \ldots$$

as the higher levels are generated.

Obviously, a whole variety of ways exists in which the discrimination processes taking place in the hierarchy can be ordered or mapped onto the record or memory system. Physically significant applications of the hierarchy will be those that preserve some interesting structure at the different levels. The first properties of the hierarchy were discovered by Parker-Rhodes (see reference (2)) who related structures at different levels in the following way. He represented any given discrimination operation as a matrix transform, and then regarded the next level as constituted of vectors of order n^2 constructed from the $n \times n$ matrices used in the transform by any invariant rule (such as by putting all columns in the matrix end to end). If we write the discrimination process

$$AB \to C,$$

for A, B, C, column vectors of order n, then there will be a class of $n \times n$ matrices X, Y ... such that

$$A[X] \to C$$

etc. for a given B. This class can be used to define an element at the new level by writing each of the $n \times n$ matrices as n^2-vectors as I have described and by forming the union of the set under the discrimination operation. Such a method of construction of a new level preserves some type of order between the levels which is potentially capable of physical interpretation, and the matrix transform provides the record system which I have already argued to be necessary in the construction of a hierarchy and which increases in size by squaring of the existing record space at each change of level.

The method of level construction which I have just described contains several arbitrary features, and in fact the level to level relationship that it defines is not easily interpretable in detail physically. In a paper now in print (see reference (6)), R. H. Atkin and I define the set of n elements at any given level as the sum of the $n+1$, n, $n-1$, ..., simplicial complexes, and the physical interpretation of the individual complexes may then be considered, and the algebra of order-preserving mappings used upon the concept of levels in a hierarchy in a natural way. However, the only detailed physical problem to be discussed in this paper—that of the origin of spin—can be treated using the matrix method which I have outlined.

It is also possible to see, in spite of the arbitrary features of the matrix method of level construction, that an upper bound to the size of each level exists and that this bound is independent of the particular choice of relation between levels. The upper bound corresponds to the case when every possible discrimination has been made, and we immediately get the sequences

Level	0	1	2	3	4	5
Maximum number of discriminable quantities		3	7	127	$\sim 10^{38}$	STOP
Cumulative sums		3	10	137	$\sim 10^{38}$	STOP
Available space in record system	2	4	16	256	STOP	

The existence of the cumulative sums, with their correspondence with the reciprocals of the characteristic coupling constants of the physical fields, gave the first indication of possible physical significance for the hierarchy. It is recognized in current physics that the coupling constants—being pure numbers—have a unique status in that their values determine the relative scale of different types of physical phenomenon. There is a history—dating especially from the work of Eddington—of development of a point of view which attaches significance to these dimensionless constants, as constraints, first upon the possible values of the natural atomic and cosmological constants of which they are ratios, and hence on all measurements. This view would be able to explain the unique status universally accorded to the dimensionless constants, but it has not made headway because of the basic difficulty that the current view of measurement precludes prior constraints on the possible range of values that any physical quantity —including these constants—can have. This difficulty does not exist in a constructive theory, and within the context of such a theory it is therefore proper to review the status of the dimensionless constants, and natural to seek theoretically determined values for them. Indeed one can see in a general way that in a constructive theory there would have to be constants having the kind of relation to measurement that the dimensionless constants have. I shall call such constants *scale-constants*.

3. Half-integral spin. The construction algorithm described in the last section was shown to generate levels of discrimination which we shall expect to be able to identify in existing physical theory. In fact the two simplest levels will be associated with spin vectors and space vectors respectively. The former consists of two and the latter of three independent (in my development—discriminable) components. (The places of time, and of relativistic invariance, in the hierarchy are very important and will be discussed later in this paper.)

The half-integral character of spin, which in several papers in this book has been regarded as the absolutely characteristic thing about the quantum-theoretic formalism, originates—I contend—in a mathematical relationship whose existence current theory has not led us to suspect. The hierarchy construction requires that operations of a given type at one level have an operation of a second type to produce the same effect at the next level. It is found that at the spin level there exist operations which have no single operation at the next level to correspond to them. They have to be replaced by two operations. There exist none that have to be replaced by three operations. It is this relationship which is usually interpreted by saying that the spin vector rotates π under a rotation through 2π of the space axes— giving half-integral spin. In current theory half-integral spin is a property of the only type of operator that would provide a two-spin state for the electron, and therefore depends on observation of spectral structure. I claim to deduce it.

I consider vectors of two elements and investigate the way in which order is preserved under discrimination operations. It is convenient to write vectors in columns and to represent the preservation of the order between the binary elements 0, 1 by writing a particular one of them in—say—the top position in the vector. The choice of which element we consider is not arbitrary once it has been decided which shall compose the null vector (and the 0 symbol was chosen for this purpose).

The discrimination operation B on the elements of a set S which was defined in section 2 may generate subsets S', S'', ..., of S which are closed under B. Thus for $x, y \in S$ we shall have then

$$x \neq y \to B(xy) \in S^i \qquad (3.1)$$
$$(\text{and} \quad B(xy) \to x \neq y \in S^i)$$

for any i. It can easily be seen that the null vector or ordered set of zero elements cannot be a member of any S^i, as otherwise the above relation will not be invariant under change of i. (This result has been assumed in stating the numbers of entities at different levels.)

There is thus a distinction built into the discrimination algebra between the two symbols 0 and 1 which is unfamiliar from ordinary linear algebra, and which is vital to the representation of spin. The simplest level of the hierarchy contains vectors of two elements, and a sufficient representation at this level of the basic asymmetry of 0 and 1 in the discrimination calculus is secured by partitioning operators into two classes: those that have 1 in the first place (top, if

TED BASTIN

we write the vectors as vertical columns) and those with 0 in the first place. Let us call the former class 'allowed', and the second 'disallowed'. We find that there exist disallowed operations which cannot be replaced by a single allowed operator. These require two allowed operations to perform the transformation that would have been produced by the given disallowed operator. There are no disallowed operators which require to be replaced by three or more allowed operators.

The operators which can be represented by two vectors, and which have a 1 in the top place provide the following transformations:

$$\underbrace{\begin{pmatrix}1\\0\end{pmatrix} \to \begin{pmatrix}0\\1\end{pmatrix}}_{\begin{pmatrix}1\\1\end{pmatrix}} \quad \underbrace{\begin{pmatrix}0\\1\end{pmatrix} \to \begin{pmatrix}1\\0\end{pmatrix}}_{\begin{pmatrix}1\\1\end{pmatrix}} \quad \underbrace{\begin{pmatrix}1\\1\end{pmatrix} \to \begin{pmatrix}0\\1\end{pmatrix}}_{\begin{pmatrix}1\\0\end{pmatrix}} \quad \underbrace{\begin{pmatrix}0\\1\end{pmatrix} \to \begin{pmatrix}1\\1\end{pmatrix}}_{\begin{pmatrix}1\\0\end{pmatrix}}. \quad (3.2)$$

The null vector is excluded.

By symmetry there are four which have a zero in the top place and which require two successive vector operations to effect them. Of these four we again reject two as being generated by the null vector, leaving two, namely:

$$\underbrace{\begin{pmatrix}1\\1\end{pmatrix} \quad \begin{pmatrix}1\\0\end{pmatrix}}_{\begin{pmatrix}0\\1\end{pmatrix}} \quad \underbrace{\begin{pmatrix}1\\0\end{pmatrix} \quad \begin{pmatrix}1\\1\end{pmatrix}}_{\begin{pmatrix}0\\1\end{pmatrix}}. \quad (3.3)$$

When we look for pairs of allowed operations that will replace these, we find only one pair in each case, namely:

$$\underbrace{\begin{pmatrix}1\\1\end{pmatrix} \quad \begin{pmatrix}0\\1\end{pmatrix} \quad \begin{pmatrix}1\\0\end{pmatrix}}_{\begin{pmatrix}1\\0\end{pmatrix} \quad \begin{pmatrix}1\\1\end{pmatrix}} \quad \underbrace{\begin{pmatrix}1\\0\end{pmatrix} \quad \begin{pmatrix}0\\1\end{pmatrix} \quad \begin{pmatrix}1\\1\end{pmatrix}}_{\begin{pmatrix}1\\1\end{pmatrix} \quad \begin{pmatrix}1\\0\end{pmatrix}}. \quad (3.4)$$

The foregoing analysis is expressed for convenience in terms of operators, but holds for the possibilities of replacement of vectors under the discrimination operation in any circumstances. We can generalize it as follows:

Let a set S be given of allowed vectors. Call this the *replacement subset*. Then

$$\mathbf{Av} = \mathbf{a} + \mathbf{v} \quad (3.5)$$

for certain vectors $\mathbf{v}$ where $\mathbf{A}$ is a matrix and $\mathbf{a}$ is a member of the replacement subset. Thus one can choose $\mathbf{v}$ and define $\mathbf{v}$ in terms of expression (3.5), and consider the question of whether $\mathbf{a}$ is a member of S. It may happen that although $A\mathbf{v} \neq \mathbf{a} + \mathbf{v}$ for any $\mathbf{a}$, S, yet there exist two (or more) vectors in S ($\mathbf{a}$, $\mathbf{b}$ say for the case of two) such that

$$A\mathbf{v} = \mathbf{a} + \mathbf{b} + \mathbf{v}. \tag{3.6}$$

This will be possible as long as $\mathbf{a} + \mathbf{b}$ does not belong to S.

One can ask: what sort of replacement subsets give rise to such behaviour? For the level of n-vectors the smallest subset which will, by repeated additions, serve for the whole lot of A is a set of n linearly independent vectors, since sums for any number of these generate the whole space. In order to give a more general idea of the relation of allowed and disallowed vectors it has been necessary to use matrix transforms which are part of the record system of the hierarchy. Its use here is only illustrative since for this very simple case a complete solution has been obtained by enumeration of cases: it does not matter that the matrix transform is only a special case of a record system.

4. Spin as a dynamical concept.

A large part of the structure of quantum mechanics has its origin in the belief that a satisfactory theory of the microscopic structure of matter needs to be expressed in terms of a mathematics which has a dynamical interpretation of a more general sort than that of the mathematics of classical mechanics, so that discrete (quantum jump) and continuous systems can be treated as special cases of the general formalism. In theories like that now being discussed this belief no longer applies, and therefore one must carefully compare the new proposals with the old to see which parts of the old have to reappear intact in the new and which were merely necessary to enable the old theory to conform to the belief which is being called into question.

It is useful to consider some observations of Pauli [3] in the paper in which he introduces the spin concept. His spin operators† satisfy

$$s_x s_y - s_y s_x = 2i s_x \quad \text{etc.} \tag{4.1}$$

where
$$s_x^2 + s_y^2 + s_z^2 = 3 \tag{4.2}$$

and where the spin operators s_i are chosen in a particular way so that, in the special case where $\mathbf{s}$ has the value $\frac{1}{2}$, the equation becomes

$$s_x s_y = -s_y s_x = i s_z \tag{4.3}$$
$$s_x^2 = s_y^2 = s_z^2 = 1.$$

---

† The s_μ are directly analogous to the vectors of (3.5) although printed as vector components. In a hierarchy, vector status is relative, depending on choice of level.

The representation of the s_i to achieve this result is

$$s_x = \begin{pmatrix} 0 & 1 \\ 1 & 0 \end{pmatrix}, \quad s_y = \begin{pmatrix} 0 & -i \\ i & 0 \end{pmatrix}, \quad s_z = \begin{pmatrix} 1 & 0 \\ 0 & -1 \end{pmatrix}.$$

Pauli remarks that in this special case equation (4.1) takes on the 'sharpened form' (4.3), so that a digital structure is achieved as a limiting case of one which can be related to conventional continuum mechanics. In achieving this result the half-integral spin appears, and the view of the digital as a limiting case makes it natural to interpret the spin vector as undergoing a rotation through an angle π when the space axes rotate through 2π, and this is indeed the conventional way of speaking.

If, however, we take the discrete structure as the primitive one, we have to re-examine the unity of the picture which Pauli presents. The success of Pauli's picture consisted in reconciling three requirements in one formalism.

(1) That the spin variables have two eigenvalues and could therefore explain the spectral observations.

(2) That the spin variables be incorporated into a quantum mechanics having the commutation relations which were at that date already established.

(3) That transformations of the spin variables be of a suitable mathematical form to be regarded as transformations of three-dimensional physical space. The resulting formalism had the half-integral spin as a consequence.

By contrast, the central advantage of the present theory is that it reverses this deductive sequence. The half-integral spin arises as a basic consequence of the method that has been used to set up an interpretable physical space of any sort—discrete in this case. Condition (1) is satisfied automatically. Condition (2) changes completely. The particular form of commutation relations are dictated by a coherent scheme of continuum mechanics which in the present approach has to be reconstructed piecemeal and which contains much that is arbitrary or historically accidental. This goes in particular for the use of the field of complex numbers as contrasted with the binary field 0, 1 in vector algebra. The one aspect of the coherence of the current mathematics of spinors and their commutation relations which seems to need more explaining away is their analogy with Jacobi's form of classical dynamics. It seems reasonable to expect that this analogy may be explicable on the basis of an extension of the de Rham theory to the topology of our discrimination calculus, but this will not be attempted in this paper.

The third of the requirements will be discussed in a later section of this paper.

5. Spin and dichotomous choice.

There have been attempts to understand the spin calculus in terms of a principle of dichotomous choice which resemble the present approach in attaching fundamental significance to binary quantities. In particular Rosenfeld [4] develops an abstract 'dichotomic mathematics' or mathematics of dichotomous choices which gives him a justification for introducing spin vectors. He says:

When we have to distinguish between two possible states of some system, such as, e.g. the two states of different charge (proton and neutron) of a nucleon, we may conveniently characterize them by the eigenvalues, say $+1$ and -1, of some quantal variable. A discrimination between the two states in question then corresponds to a 'measurement' of the variable, such a measurement yielding the answer $+1$ or -1. In a matrix representation, such a *dichotomic variable* will appear as a Hermitean matrix with 2 rows and 2 columns, and will satisfy the equation

$$\tau^2 = 1.$$

From these conditions, it is easily deduced that the most general form for τ is

$$\tau = \begin{pmatrix} a & \sqrt{(1-a^2)}\,e^{-i\phi} \\ \sqrt{(1-a^2)}\,e^{i\phi} & -a \end{pmatrix}.$$

From this form, an interpretation of the dichotomic variables in terms of spatial directions is straightforward. Rosenfeld follows the method of establishing an interpretation for his new structure and then identifying that structure with existing physical concepts at each stage in its development, just so far as that identification is justified by the formal correspondence of the new structure with current theory. This is also my method. The difficulty with this method in either case is that you can never call a halt and say you have arrived. I am prepared to swallow this pill: Rosenfeld is not. In Rosenfeld's case a logical development from the idea of dichotomous choice is made to revert suddenly to the quite different ideas on which current physics is ultimately based, and you cannot do this. Indeed in general it is not possible to get support for a theory partly from each of two unrelated sets of ideas.†

---

† Even in law, where I am allowed to plead either that I did not kill X or that if I did, there were extenuating circumstances, neither plea is held to be strengthened by the existence of the other.

Rosenfeld attempts to avoid this trouble by clearly separating his dichotomic choice calculus from its application.

The correspondence here discussed between dichotomic variables and spatial directions is, in the general case, purely symbolical. But it acquires a real significance in the theory of spin, or intrinsic angular momentum, of the electron or nucleon. For the establishment of this concept the existence of a set of dichotomic variables forming a vector, together with the relations between them, is of essential importance.

The attempt does not really work though: the understanding of spin in terms of dichotomous choice *depends* upon the discreteness of the variables which represent that choice, and in this case a continuous application of a discrete calculus precludes the basis of understanding of that calculus.

My proposal regarding Rosenfeld's analysis of spin is not to reject it itself but to reject the possibility of suddenly jumping from it into the use of the modified current concepts. Such a proposal implies a programme along similar lines to the one I am describing in this paper.

6. Lorentz invariance in a discrete theory.

If we accept the identification of spin with certain operators in the simplest levels of a discrimination hierarchy, then the identification of the three discriminable 2-vectors with space directions follows, This identification at first sight looks unnatural, though, as Rosenfeld also points out, remarkable. The feeling of unnaturalness arises, in fact, because one tends to think that one cannot use the language of dynamics at all without using all of it, and in particular if one speaks of the directions of space then the whole possibility of considering indefinitely large sets of points in that space must already exist. Instead, we have the following physical picture: any physical phenomenon is defined by some special restriction on, or structuring of, the discrimination process at some level, and if one wishes to know more detail about that phenomenon one does experiments and attempts to incorporate the knowledge one has gained by identifying more structure at that level. This is not necessarily possible—a fact which was first forced on our attention by the very breakdown of classical physics which led to quantum theory. In these circumstances one has to change to a new level with more elements. There are two effects that this change may have: first, a possibility of making new distinctions appears, and secondly, distinctions hitherto made may vanish.

The converse process of moving to simpler levels is more familiar. Any newly discovered detail in any physical phenomenon is always ultimately analysed—if it can be analysed at all—in terms of the concepts of mechanics and electrodynamics and these concepts derive their non-metrical characteristics from the algebraic structure of the simple hierarchy levels, to which one has to return recursively in any application of the hierarchy.

Since we reject attempts like that of Rosenfeld to reconcile conventional space-time with spin seen as an abstract principle of dichotomy it seems we must sacrifice some important insights. In particular there seems no place for Dirac's derivation of the spin vectors from a linear four-dimensional form of the wave equation, which is usually held to demonstrate a connexion between relativistic invariance and the quantum world.

I argue that no insights are sacrificed, but that on the contrary certain puzzles could be removed from the area of overlap of relativity and quantum theory by adopting the idea of physical space as a constructive space using the hierarchy model. My argument consists of the following steps.

(1) The connexion between 4-component spin matrices and the Lorentz group is accidental and limited to the common appearance of the number 4. The familiar treatment of Dirac [5] who showed that a 4-component wave equation with linear operators α_μ afforded a satisfactory representation of the Pauli spin variables, can be related to my present approach: Dirac's 4×4 matrices ρ_μ, which define a representation of the spin vectors in terms of the α_μ through

$$\alpha_1 = \rho_1 \sigma_1, \quad \alpha_2 = \rho_1 \sigma_2, \quad \alpha_3 = \rho_1 \sigma_3, \quad \alpha_\mu = \rho_3$$

can be identified with the 4×4 matrix forms which I introduced earlier as a possible form of the necessary record system.

(2) There is no need to attempt to identify the dimension number of space-time with the order of 4-vectors in the spin theory, since in the discrimination algebra both appear, but in *a quite separate manner*.

(3) If Lorentz invariance has not got to be found a place in quantum theory then we no longer have to struggle with the extremely puzzling fact that the operational basis of relativity theory and hence the experimental reasons for imagining time as a dimension analogous to the space dimensions are so remote from those of quantum theory with its predominantly local, laboratory scale experimentation.

It is possible, consistently with the discrete approach to space that I am proposing, to regard the assimilation of time to space that underlies relativity theory as a mathematical device to procure a limiting velocity (namely the velocity of light) and hence to give a primary place to the dimensionless scale-constants (see section 2 of this paper) without a change of physical concepts at the most basic level. A brief account of this possibility has been given[2] and a fuller development of it is being written by Atkin and myself[6]. I shall not discuss it further now, its place in my present argument being only to indicate how to make the insights of relativity theory fit into the picture.

The picture in fact provided by the hierarchy constitutes a dynamic view of the acquisition of knowledge about any physical process, and it is natural to identify time with the sequence of stages through which the hierarchy passes. Obviously, it is important to keep this primary meaning for the word 'time' quite separate from the sophisticated concept of co-ordinate time. The reason it is so vital clearly to distinguish these two meanings is that very important deductions can be made from the theory as consequences of the conditions that have to be satisfied in advancing from the one time concept to the other. The treatment of spin which has been the subject of this paper can be seen as such a consequence.

The physical picture which emerges from my approach lacks the permanent objective background which the physicist expects to exist automatically. Each new investigation has to define its own starting point and construct its own spatial distribution of objects as it develops ('in time', if one wishes so to speak). Such a theory may have a very real advantage in dealing with high energy quantum systems where, for example, we are strongly impelled to think in terms of time-reversal—a concept which is practically inconceivable against a conventional invariant space-time background. In a theory which does not presuppose an invariant space-time background there will be an initial problem of constructing enough of such a background for any application that is being considered. One will have to ensure that one can fit the bits together. This requirement explains the importance of the scale constants in my approach. They are invariant for all applications of the theory and will therefore assume initial importance in interpretation, by contrast with their position in current theory where they are remote from natural interpretation. Further developments of this theory will require the discovery of quantities of an intermediate status which have some degree of invariance so as to make it plausible to interpret them as physical quantities. Then the conditions for them to be invariant prescribe the nature of their interpretation: they will be invariant in a certain context and that context will determine their interpretation.

7. Computing methods. The discrimination calculus results in a hierarchy in which at each level the points are constructed from structures (ordered subsets) at the previous level. One does not, however, begin afresh at each level because of the record system for relating structures at the different levels. This record system is easily considered as a digital computer memory, and since many of the properties of a discrimination hierarchy appear in a natural form when thought of as the operations of a computing system (this is specially true of the representation of physical time in a hierarchy) I conclude this paper with an idealized computer program to illustrate the use of the record system in relating levels.

The program is written in a much simplified form of TRAC (I begin by stating TRAC in four primitive terms and operations). TRAC itself is a high level programming language[7], [8] which is implemented and used at Cambridge Language Research Unit because of its logically primitive character and its applicability for purposes like the one I have now in hand.

PRIMITIVE TERMS AND OPERATIONS

1. *String*: an ordered set of characters or ciphers.
2. *Store*: set of ordered pairs each consisting of one string with one name. At any given time there is a one to one correspondence of strings and names.
3. *Call*: operation of drawing a string from store by specifying its name. The string also remains in store.
4. *Define*: operation of attaching a new string to a given name (and therefore of placing the string in store). A string defined with a given name replaces any string already in the store of that name.

FORMAL RULES

1. There exists a set $I = I(A, B, \ldots$ to h terms) of binary strings (strings consisting entirely of the binary units 0, 1), of equal length l, in store at a given time.
2. The names $a, b, \ldots$ to h terms, of the strings in I are binary strings all of length n.
3. A generating process is defined for constructing members of I. If P and Q (P and Q in I) are called, then a new string R is defined such that for all P, Q there exists a unique R.
4. R is of the same length as P and Q.

5. If and only if P has the same binary unit in each given place as Q has in that place, R is the null string consisting of zeros in each place.

6. A sequential process is begun by calling two non-null strings P, Q. We write the generation of R, $D(PQ) \to R$. (In the following rules 7, 8, 9 we specify the selection of the next string.)

7. The name $N(qr)$ of length l^2 is formed from the names q, r in such a way that each digit i_q in q and each digit j_r in r determines the choice of some digit ij in N, that way being independent of i and of j.

8. $N(qr)$ is the name of a member, P', of a second set of strings I'.

9. The set I' has as its elements strings which are ordered subsets of the $2^l - 1$ non-null elements of I, and which obey the same principle of generation as the elements of I.

10. The name of the new string s' in I is an ordered set of n digits of P' selected according to some consistent rule which I do not here specify.

The foregoing set of rules contains arbitrary elements in rules 3, 7, 10 which can be eliminated by completing the algorithms by randomizing over the choices left unspecified. A combination of rules 3 and 5 is sufficient to specify the generation operation as discrimination as defined in this paper and in the case of rule 3 the non-specificity is a formal necessity. The arbitrariness in rules 7 and 10 is real, however, and the actual choice of specific rules will determine the application of the program in detail. Nevertheless all the deductions made in this paper would remain true if the unspecified choices in rules 7 and 10 were made at random.

REFERENCES

(1) Brouwer, L. E. J. Points and spaces, *Can. J. Math.* (1954) **6**, 1.

(2) Bastin, E. W. On the origin of the scale-constants of physics, *Studia Philosophica Gandensia* (1966) **4**, 77.

(3) Pauli, W. Zur Quantenmechanik des magnetischen Elektrons, *Z. Phys.* (1927) **43**, 607.

(4) Rosenfeld, L. *Nuclear Forces* (1949), 40.

(5) Dirac, P. A. M. *Quantum Mechanics* (Third edition; Oxford, 1947), 254.

(6) Atkin, R. H. and Bastin, E. W. *Int. Journ. Theor. Phys.* (in print, 1970).

(7) Mooers, C. N. T.R.A.C.—A text-handling language, Paper presented at 20th Nat. Conf. A.C.M., Cleveland, Ohio (Aug. 1965).

(8) Mooers, C.N. T.R.A.C.—A procedure-describing language for the reactive typewriter, *Comm. A.C.M.* IX (1966) **3**, 215.

VI
PHILOSOPHICAL PAPERS

In this, the final part of the book, papers appear which contain general comment on the quantum problem. They may or may not be philosophical in the sense that they approach that problem from the point of view of the professional philosopher. Most of the discussion at the colloquium was on subjects which require technical philosophical expertise if any degree of precision in the treatment of them is to be achieved. Comparatively few of the participants possessed this expertise,† and the discussions on the whole were too diffuse in form to be effective in print. We do however choose as our concluding text a discussion which took place between von Weizsäcker (philosopher and physicist) and Linney (philosopher and logician) on the relevance of quantum-theoretical epistemology to the traditional philosophical problems of knowledge and sense-data. Von Weizsäcker's paper itself exhibits a not completely resolved conflict between the epistemology he presents as that of orthodox quantum theory and a form of finitism which with Drieschner he develops from a 3-tense logic. The tentative theoretical extensions of physical theory—particularly into cosmological fields—which conclude his paper, owe their stimulus to this finitist position.

Bunge describes his approach as 'realist'. This realism did not make much impact on the discussions—probably because it assumes that quantum theory can be accepted as a technical and completely mathematical theory which can be accepted or rejected independently of any conceptual difficulties. No one else present was prepared to make this assumption.

Prima facie the information concept in the sense that information theory has made familiar seems to constitute one of the fruitful approaches to the quantum problem. As it turned out, Rothstein was left to fight a lone battle on this front. In his paper Rothstein presents the case for thinking that the puzzles of quantum theory vanish in

---

† To find philosophical sophistication and scientific invention in one mind is now regrettably rare, and that rarity was part of the setting of this colloquium, as well as the problem facing it. Even to get a lot of specialists in the philosophy of quantum theory was no solution, for such specialists easily give the impression that they have a strong vested interest in the difficulties of quantum theory remaining more or less stationary.

a context in which exchange of information is understood to be the basis for measurement. Rothstein proposes a thermodynamic explanatory background for his use of 'information'.

Information is quantified in binary units, and it is therefore simple to argue that—for example—the record of the result of a measurement must ultimately be contained in discrete changes in one or more atomic structures, and therefore the wave-function must have collapsed. Of course such an argument is facile unless one can provide a reason why there should be naturally occurring units of information. Rothstein argues that there cannot be indefinitely exact localization of the representative point in the phase space of a quantum system because this would require infinite entropy increase. Hence there must be a discrete grid which is available to define a set of units of information. Such a view clearly presupposes a very realist interpretation of thermodynamic potentials—entropy having an existence in its own right and with a background theory which depends upon its being conserved. Most colloquium members were not prepared to grant such a theory and would have regarded entropy as statistically defined and as having no epiphenomenal meaning. For them Rothstein's position would not be tenable. Rothstein however envisaged a biological background for the thermodynamic potentials, and, though he did not make detailed suggestions, believes that such a background could in principle give the entropy concept a status outside its definition in physics.

The bearings of biology upon quantum theory were further pursued in the last paper of this book—by Pattee. Pattee did not seek to find room for the processes responsible for life in the gaps in our understanding of the quantal picture (a position which has its supporters at present but which was not put forward at this colloquium). He preferred to think that if we look critically at the basic logic that is imposed on us by having to include the code for molecular construction within the molecule, we shall also find guidelines for the measurement problem of quantum theory itself.

This proposal was not seriously discussed by the colloquium. Probably it was felt to be introducing material too different from what was being introduced by other speakers. It seems quite likely however that studies of the nature of codes will have to be considered by later meetings of the 'Quantum Theory and Beyond' sort.

THE UNITY OF PHYSICS†

C. F. VON WEIZSÄCKER

1. *A list is presented of unsolved conceptual problems in the interlinkage of five basic theories: relativity, quantum theory, elementary particle theory, thermodynamics, cosmology. The problems might be solved by a unified physics. The epistemological hypothesis is formulated that this unified theory will not presuppose more than the very preconditions of experience.*

Is there a way beyond a mere acceptance of quantum theory as an empirically well established set of laws? We should like better to understand either its necessity or its possible future alterations. For both purposes it may be useful to ask what further progress in physics can be expected with any plausibility.

Heisenberg has described the past progress of theoretical physics as a series of distinct 'closed theories' ('abgeschlossene Theorien'). In sharp contrast to the smooth piling up of empirical data and of their explanation by existing well-established theories, the basic theories seem to advance in rare great steps or jumps. The decisive step forward is certainly historically prepared, but in many cases not with the accompanying feeling of growing clarity but rather with the increasing awareness of some unresolved riddles. This historical phenomenon is most clearly seen in the years preceding the formulation of special relativity and of quantum mechanics. A closed theory is generally characterized by an intrinsic simplicity, but no methodology of science has so far been able clearly to define what we mean by 'simplicity' in such a statement. In any case closed theories show a remarkable ability to answer those questions that can be clearly formulated within their own conceptual framework, and to give its followers the feeling that the questions which cannot be so formulated may be altogether meaningless questions. In the historical sequence of closed theories the later ones usually reduce their predecessors to some 'limited' or 'relative truth', assigning them the role of approximations or limiting cases. Thus we have learnt to speak of the field of applicability of a theory, the limits of which are not known in the beginning and are clearly defined only by the later theories.

† 'The Copenhagen Interpretation' published in Part II of this book, was originally the first part of this paper.

It is one of the most important tasks for an epistemology of science to explain why theoretical progress should have this particular form. I venture the hypothesis that any good, that means widely applicable, theory will be deducible from a very small set of basic assumptions. A possible description of the feelings of good physicists about a 'good theory' is that it does not admit of minor improvements. If a good theory follows from very few qualitative postulates this would be expected; the only possible changes in such a theory are changes in the basic postulates, which will be felt to be 'great' changes. This hypothesis presupposes that a successful theory in physics stands under very severe constraints. Logical and mathematical consistency is only one of them; semantical consistency may turn out to be the most severe one.

We can fill this methodological scheme with content by looking at the present situation of theoretical physics. Perhaps one can organize our present knowledge and our present expectations by saying that five interlinking fundamental theories either exist or are sought for; the question, how they interlink, will be our main problem. They are:

1. A theory of space-time structure (special or perhaps general relativity).

2. A general mechanics (quantum theory).

3. A theory of the possible species of objects (elementary particle theory).

4. A theory of irreversibility (statistical thermodynamics).

5. A theory of the totality of physical objects (cosmology).

This list excludes theories of special objects like nuclei, atoms, molecules, wave fields, stars etc. which, at least in principle, can be deduced from the fundamental theories. We are inclined today to consider nos. 1, 2, and 4 as more or less final, while much work is being done in order to find no. 3 and perhaps no. 5.

In contradistinction to earlier theories these present ones no longer depend on particular fields of experience like sense-data (optics, acoustics, etc.), moving bodies (classical mechanics), or fields of force (electrodynamics). They seem to arrange themselves like parts of a systematic unity of physics which is as yet dimly seen. We may try to express the principle of this unity by saying: There are objects in space and time. Hence an account of space and time must be given (no. 1). 'Being in space and time' means for an object that it can move. Hence there is a set of general laws that govern the motion of all possible objects (no. 2). All objects can be classified in more or less distinct species; were it not so, no general concepts of objects could be formed and there would be no science. Hence there must be a theory telling what species of objects are possible (no. 3). This theory describes objects as composites of more elementary objects. The composition can be described in detail, leading to the higher species (atoms, molecules, etc.). It can also be described in a statistical manner (no. 4). All objects of which we know somehow interact, or else we would not know about them. Hence some theory about all existing objects may be needed (no. 5).

This is only a preliminary account of a possible unity of physics. Its shortcomings are seen when we more closely analyse the interlinkage of the theories and the problems connected with the concepts I used in the description.

(1) The space-time structure interlinks with all four other fields in a rather puzzling way. (1 and 2): Quantum theory in its only achieved form presupposes time but does not presuppose space; it only describes the manifold of the possible states of any object by the highly abstract concept of Hilbert space. (1 and 3): According to general relativity, the space-time structure is described by gravitation, which on the other hand seems to be a field that one would like to deduce from elementary particle theory. (1 and 4): Thermodynamics deals not only with statistics but also with irreversibility, which seems to be a feature of time. (1 and 5): Cosmological topology is itself a theory of space-time structure, this in turn perhaps depending on the gravitational field.

(2) The concept of a physical object (or 'system') which is presupposed by quantum theory contains many problems. An object seems to be an object for a subject (for an observer of phenomena); the fact that the observer is himself part of the objective world is accepted as fundamental in the Copenhagen interpretation but it is not objectively described by any of our five theories. If we leave this question aside as too philosopical, we still encounter conceptual problems in the use of the idea of an object. The concept of an isolated object is only an approximation, and according to quantum theory a very bad one. The rule of composition states that the Hilbert space of a composite object is the Kronecker product of the Hilbert spaces of its parts. This implies that only a set of measure zero out of the states of a composite object can be described by assigning definite states to its parts. Yet we formally rely on the description of the largest objects we consider in any given situation as being isolated. (2 and 5): It seems quite speculative to describe the universe as a whole as one quantum-mechanical object that might have a defined

state vector. (2 and 4): If my analysis of the Copenhagen interpretation was correct, we rely for the description of actual quantum states on measuring instruments for whose description we explicitly renounce the precise knowledge of their quantum states as we describe their functioning as irreversible; this seems to mean that we draw on thermodynamics in order to give a meaning to quantum theory.

Other puzzles arise if we analyse our usual way of expression (which I tried to condense in my above description of the connection between theories 2, 3 and 5). (2 and 3): I described quantum theory as stating the general laws of motion for all possible objects, and elementary particle theory as hoping to describe all possible *species* of objects. What does this distinction mean? Either quantum theory and elementary particle theory will in the end turn out to be coextensive and then, probably, identical; or objects will be thought of which would be possible according to general quantum theory but which are excluded by the additional information of elementary particle physics. The second alternative expresses the conventional view. But then the quantum theory of the 'rejected' objects turns out to be physically meaningless; should we retain it at all? The distinction between universal laws and individual objects is quite meaningful, as long as the presence of these particular objects and no others is taken to be contingent, i.e. not to follow from a universal law; but if there are laws to confine possible objects to certain classes, what is the empirical meaning of more general laws? Hence I would tend to consider the first alternative as a real possibility. It might turn out that quantum theory can be semantically consistent only if the Hilbert space is described by using a system of basis vectors that corresponds to 'possible species of objects' in the sense of elementary particle theory. I shall resume the question in section 4.

(1, 2, 3 and 5): Our usual way of speaking of cosmological models is perhaps even more conspicuously awkward. We first formulate a general equation, e.g. Einstein's equation of the metrical field. Then we take out one of its solutions and state: this one (hypothetically) describes the whole world. What then is the empirical meaning of the other solutions and hence of the whole equation? Certainly the equation applies meaningfully to the various situations within the world. Their initial and boundary conditions specify the particular contingent situation to be described by an appropriate solution of the equation. We can subsume varying contingent cases under a general law. But is it meaningful to subsume the unique totality of being under a 'general' law? We are here approaching the ancient puzzle of Leibniz's concept of 'possible worlds' and the more recent riddle of Mach's principle.

(1, 3 and 5): W. Thirring has shown that a Lorentz-invariant field theory of gravitation will admit of a gauge transformation which finally leaves us with observable space-time coordinates obeying a Riemann metric and with a gravitational field obeying Einstein's equation. Thus the connection of the observable space-structure with gravitation is explained or rediscovered, and Einstein's approach of beginning with a Riemannian metric is justified. But what cannot be changed by the gauge transformation is the topology of space-time at large, that is, cosmology. If we start in a Minkowski space, cosmic space will remain open in the final description; if we wish to get a closed Einstein universe we must start out with a field theory of gravitation in some closed space. This fact indicates that cosmology might not be a consequence of gravitation. Cosmic topology might be determined otherwise while matter and gravitation (curvature of space) would be left to adjust themselves to the given boundary condition.

(2 and 4): I have already alluded to a difficulty in the connection of quantum theory and thermodynamics. We usually think of irreversibility as some sort of secondary effect, superimposed on essentially reversible basic laws, and to be explained by classical statistics, that means by our lack of knowledge. On the other hand, in my paper in Part II of this book I used irreversibility as a precondition of measurements and hence of the semantics of quantum theory (or any other theory which rests on observation). (4): In order to understand this apparently vicious circle we must first remove a flaw in the usual description of irreversibility. Boltzmann's H-theorem proves that a closed system which at a time $t = t_0$ is in a state of non-maximal entropy will with great probability be in a state of higher entropy at a time $t > t_0$, that is at a time which at t_0 is still in the future. That proves the second law for the future, but not for the past. If we apply the very same consideration to a time $t < t_0$, we will of course find that with the same probability the system will be found to be at a higher entropy level at t than at t_0. Yet we know the second law empirically, that means from the past. Hence this naive application of the H-theorem does not prove the *known* second law at all; it rather contradicts it. The apparent paradox is removed by the remark that the concept of probability in the particular meaning in which it is used here can only be applied to the future but not to the past. The future is 'possible', i.e. it is essentially unknown, and we can ascribe

objective probabilities to its events. The past is 'factual', it can be known in principle and is known in many cases; the probability of a past event only means the subjective lack of knowledge. I shall return to the logical structure of this argument in section 2. If we accept this solution of the paradox we can proceed to proving the second law for the past by remarking that every past time once was the present time. Then the concept of probability could be meaningfully applied to what then was still in the future, leading to the correct prediction that entropy would increase.

(4 and 5): Boltzmann tried to avoid this explicit use of the concepts of present, past and future by proposing that the infinite universe is on the average in a state of statistical equilibrium, containing regions in space and time where there are fluctuations. He thinks that we live in a large fluctuation, that only in it are living organisms possible, and that they will always 'measure time in the direction of increasing entropy'. Apart from feeling that 'measuring time in a direction' is a meaningless phrase, since all measurements are made *in* time, I think the argument breaks down due to a remark of P. and T. Ehrenfest. They show that any non-maximal entropy value of a system which is on the average in statistical equilibrium is with overwhelming probability not on a slope of a larger fluctuation but just a minimum value, that is a peak of a fluctuation. (This expression is precise for discrete entropy-changes; it can be adapted for continuous changes.) This statement is in fact a necessary precondition of the H-theorem. Applied to Boltzmann's idea it means that if we live in a fluctuation it is overwhelmingly more probable that the present state with all its apparent traces, fossils, and documents of the apparent past is just an extreme fluctuation than that such a past, implying even lower entropy values, has really existed. That, I think, means a *reductio ad absurdum* of Boltzmann's proposal, and thus an indirect justification of the use of 'tenses' in physics.

(1 and 4): These considerations would suggest that neither is the second law fundamental in itself nor can it be deduced from reversible laws only, but that it derives from a more fundamental structure of time which we express in speaking of present, past, and future. This structure then will be a presupposition of a meaningful use of the terms of thermodynamics as well as of quantum theory. This time structure is not described by special relativity but is not contradicted by it either, for the distinction between past and future events for a given observer is Lorentz-invariant.

Trying to put together the pieces left in our hands by this critical analysis, we may guess that there is a more fundamental unity behind the existing parts of the five theories. This unity may just be hidden by the naive way in which we still use unclear but necessary concepts like space, time, object, measurement, probability, universe. If this unity will again be expressed by a 'closed theory' we may wonder what will be its basic postulates. Thus we return to epistemology for a moment.

Epistemology has a tendency to limp behind the actual development of science. Thus one of the much discussed problems is how we can establish empirical laws. That was a meaningful question at a stage of physics in which many apparently independent laws were brought forward as hypotheses and refuted or more or less validated by empirical evidence, as e.g. the laws of Boyle and Mariotte, of Coulomb, of Ampère and others. Thus it was quite a meaningful question in the days of David Hume and Immanuel Kant. But in present-day physics these laws are no longer independent. We accept them as necessary once we have accepted the basic theories from which they follow. Thus today the only meaningful question is how we establish basic theories. If the unity of present-day physics should in the end turn out to be embodied in one closed theory, the only remaining question would be how to establish (to derive, to confirm, or whatever expression you prefer) the basic postulates of that theory.

Now the epistemological problem whether one can establish a strict law by particular experience has, I think, been answered in the negative. This was known to Plato as well as to Hume. Popper correctly points out that a universal law cannot be verified by an enumeration of empirical instances. I think he is mistaken in believing that it can at least be falsified by a counter-instance. For an empirical phenomenon to be a counter-instance against a law we always need some theoretical interpretation of the observed phenomena, that means some laws. If these laws cannot be strictly verified it will be impossible to use them for a strict falsification of another law.

I can think of just one justification of general laws in view of experience, that is of a justification which is neither dogmatically aprioristic nor just a *petitio principii*. It is given by Kant's idea that general laws formulate the conditions under which experience is possible. Such laws will not be known by us to be necessary in themselves, for we do not know that it is necessary that experience should be possible. But they will hold to the extent that experience is possible, and hence they will have to be admitted by anyone who is prepared to accept empirical evidence. The question arises whether the basic assumptions of a unified physics might just be those assumptions that are necessary if there is to be experience.

2. *The most general presupposition of experience is time. Its structure, as expressed by the words present, past and future, is analysed in a logic of temporal propositions (tense-logic), which provides the conceptual frame for quantum logic and for the theory of objective probability.*

A possible definition of experience may be that it means to learn from the past for the future. Any experience I now possess is certainly past experience; any use I now can still hope to make of my experience is certainly a future use. In a more refined way one may say that science sets up laws which seem to agree with past experience, and which are tested by predicting future events and by comparing the prediction with the event when the event is no longer a possible future event but a present one. In this sense time is a presupposition of experience; whoever accepts experience understands the meaning of words like present, past, and future. Grammatically speaking he understands the meaning of tenses. Yet language is not the decisive factor here; if he speaks a language that does not express tenses in a simple manner he still will understand the difference between yesterday and tomorrow, between what has happened and what will (or will not) happen.

I shall try to find out how much is already implied by the postulate that physics, if it is to be an empirical science, must refer to time. I shall formulate the postulate in the most general way by indicating some basic ideas of a logic of temporal propositions, or a tense-logic. This logic will turn out to be closely connected with what has been called quantum logic.

J. v. Neumann first proposed to compare projection operators in Hilbert space with propositions, and their eigenvalues 1 and 0 with the truth-values 'true' and 'false'. Thus in an eigenstate of the operator the corresponding proposition has the corresponding truth-value, in other states it does not have a well-defined truth-value, but just a 'probability to turn out true'. This logic is non-classical. The corresponding lattice of propositions is not Boolean, rather it is isomorphic with the lattice of the subspaces of Hilbert-space, that means with a projective geometry.

J. v. Neumann had good reasons for using this logical language. Probability is a fundamental concept in quantum theory. Asking what laws quantum-mechanical probabilities obey one will first ask whether they conform to Kolmogoroff's axioms of probability. These axioms introduce probability as a real function defined on a Boolean lattice of what is called possible events. That events should form a Boolean lattice follows from classical logic, since two events can be connected by 'and' and 'or'. In quantum theory what is changed does not seem to be the set of ensuing axioms but the lattice to which they are applied; this seems to indicate a change in the underlying logic.

Yet there is an apparently strong argument against the possibility of a quantum logic. This logic seems to be derived from quantum theory, that is from particular experience. Quantum theory, however, has been built up using classical mathematics; experience itself does not seem possible without using logic and mathematics. Mathematics rests on classical (or possibly intuitionist) logic. So how can quantum theory justify a logic which differs from the logic on which it is founded? To this argument there are two successive replies, one of which is a defence while the other is a counterattack.

The defence says that classical logic (under which I here subsume, for the sake of brevity, intuitionist logic) refers to one type of propositions, quantum logic to another type. Classical logic is to be accepted for 'timeless' statements like 'two times three equals six'. Quantum logic refers to 'contingent' statements like 'there is an electron at x'. Timeless statements, if true or false, are true or false 'forever', without respect to time. Contingent statements, in the phrasing that refers to the present, can be true now, false at another time. It can at least be considered possible that these different types of statements will obey different logical laws.

The counterattack says that tense-logic, of which quantum logic is just a special formulation, is not a result of particular experience but a presupposition of all experience. Quantum theory has only made us aware of logical distinctions we were permitted to neglect in classical science. This is analogous to the contention of the intuitionist, that while Brouwer was historically later than Russell, Russell's paradox just made it clear that one ought always to have thought more carefully about infinite sets. But this counterattack puts us under an obligation actually to show what logic is appropriate for contingent statements. I cannot fully develop the argument in this paper, but I shall mention the main points.†

---

† This tense-logic is formally different from the existing systems of which Prior has given a thoroughgoing account.

I shall use a 'reflective' method which corresponds to Bochenski's statement 'the true logic is metalogic'. Not any formal system that resembles usual logic is to be called a logic, but only a system that formulates rules according to which we really argue, when we try really to convince each other. In particular I follow the method of Lorenzen by which he develops the logical laws as rules according to which one can win a discussion game in a fair dialogue.

As an example I give Lorenzen's way of proving the implicative law of identity, $A \to A$. If one partner in the dialogue, called the proponent, is prepared to defend a proposition of the form $A \to B$ (A implies B), he in effect offers the opponent the following game: If the opponent is able to prove A, then the proponent will prove B. If the opponent cannot or does not prove A, then the proponent is not under an obligation to prove anything. The game is only meaningful if the partners agree beforehand about what they are prepared to accept as a proof for a 'primitive statement' of the sort for which A and B stand as symbols. If A and B are arithmetical propositions, a proof may be an arithmetical construction; if they are well-formed expressions of a formal system, a proof may be their derivation from some accepted initial expressions. Lorenzen's view is that the validity of logical laws will not depend on what sort of proof is agreed upon, but only on the existence of such an agreement. Now assume the proponent proposes to defend the general law $A \to A$. If the law is to be general he must permit the opponent to insert for A whatever proposition he likes. Let him insert some particular proposition a (e.g. '$2 \times 3 = 6$'). He is asked by the proponent to prove a. Now the way divides. Either he does not give a proof. Then the proponent is not obliged to do anything. Or the opponent gives a proof for a. Then the proponent is obliged to prove what follows by inserting the same a for A; that means he is obliged to prove a. This he does in taking over the opponent's proof. Thus he wins the game in every single instance, and, what is more, he knows in advance that he will always win since he knows a winning strategy. This strategy (to take over the opponent's proof) is the 'meaning' of the law $A \to A$.

Now let us repeat the game for contingent propositions referring to the present. For A we insert m, which is meant as an abbreviation for 'the moon is shining'. How can we prove a statement about the present? The simplest and fundamental proof is by inspection: 'I show you the moon, please look at it.' The game now seems to follow the same pattern as before. The proponent offers to defend $A \to A$. The opponent inserts m. Asked to prove m he says: 'Here look at the moon, just above the horizon!' The proponent admits the proof. Asked now to prove m, he says 'Here look at the moon, just above the horizon!'. The opponent has to admit the proof which means to admit defeat.

Yet the proponent must in this example be careful to react fast enough. Else the opponent who just before showed him the moon may deny the validity of the second proof: the moon has set in the meantime. If such things can happen, it is not self-evident that all laws of classical logic can be reproduced for contingent statements about the present. Mittelstaedt has in fact shown that the law $A \rightarrow (B \rightarrow A)$ for precisely such reasons does not hold in quantum logic. In classical logic the law is defended like this: let the opponent insert some p for A. Let him prove p. Now the proponent must defend $B \rightarrow p$. Let the opponent insert any q for B. Let him prove q. Now the proponent must prove p. He resumes the proof for p given by the opponent, and wins. In quantum theory let p mean 'this electron has the momentum p' and q: 'the same electron has the position q'. Measurements of the respective quantities are admitted as proofs. The opponent first measures the electron's momentum and finds p, then he measures its position and finds q. Now the proponent resumes the momentum measurement, but alas, he does not find p again.

This way of arguing must provoke severe objections. One will say that the whole mode of expression of statements referring to the present is imprecise because the present is not always the same time; thus statements are identified which really are different. If in the Mittelstaedt example we denote the times of the three successive measurements by 1, 2, 3 respectively, the first statement on momentum will read p_1: 'at the time 1 the momentum is p', while the last one will read p_3: 'at the time 3 the momentum is p'. p_1 and p_3 now are statements about 'objective time moments' which can be considered to be timelessly true or false. There is no necessity why p_3 should be true if p_1 is true.

It is decisive for tense-logic to stay firm against this objection. The counter-objection is that we have to rely on *some* statements about the present in order to be able to prove any statements at all, be they timeless or contingent, about objective time-moments or in tense form. A statement about an objective time-moment t_0 can be proved by inspection just once in history, namely at the time t_0. No proof by inspection for such a statement can ever be given at any other time. How can logical laws about them be justified then? We may prove p_1

at time 1 by inspection. If we are to prove p_1 once more (in contra-distinction to proving p at time 3, i.e. p_3), we must rely on documents or traces of the past. If such traces exist we will have to recognize them at the given time by inspection, or by having a trace of the trace. This means: statements about past objective time moments can meaningfully be made only if statements about the present can meaningfully be made. The possibility of the formulation of contingent statements referring to objective time moments, which traditional logic would accept, rests on the possibility of their formulation referring to the present, which traditional logic would like to eliminate. All that amounts to the truism that whenever 'I am speaking' is true, 'I am speaking now' is equally true; I cannot see a phenomenon, or utter a statement, but 'now'. I would be prepared to accept the view that here we are facing the deep mystery of time, but I would add that unless we really face it we will not even understand what we mean by logic, let alone physics.

What about statements about the future ('the moon will be shining tomorrow')? Statements about the past can rely on traces. Of the future there are no traces. I made use of this difference between past and future in discussing the second law of thermodynamics in section 1. I think we have to accept the difference as a fundamental structure of time. However, some proof of consistency can be given. We can express the second law by saying that improbable states are followed by more probable ones and preceded by more improbable ones. A trace is an improbable state. It implies even more improbable states, that is much information, about the past; but it implies more probable states, that is very little information, about the future. This sheet of papers contains letters; it is a document of the improbable event of some particular human thought in the past. Its future, however, is quite uncertain: it may be forgotten in a library, burnt by fire or gone with the wind. Thus the second law both presupposes and explains that there are traces of the past and none of the future.

I am prepared to call a statement true if it can in principle be proved, whether by direct inspection (present), by inspection of a trace (past), or by mathematical or similar arguments (timeless). But I would propose not to use the values 'true' and 'false' for statements about the future. This is, I suspect, what Aristotle had in mind when he wrote his chapter 9 of *De Interpretatione*. A statement about the future can be proved or disproved by inspection only when it is no longer a statement about the future. Statements about the future can, however, meaningfully be called 'necessary', 'possible', 'impossible', 'probable with probability p' and so on; 'modalities', as logicians say, can be applied to them. 'It is necessary that the moon will be shining tomorrow' means that my contingent knowledge of the present or the past together with my knowledge of the laws of nature implies that the moon will be shining. If A stands for my knowledge, and M for the prediction concerning the moon, what I can maintain is 'necessarily if A then M'. By an admissible rule of detachment this is transformed into 'necessarily M'. Yet this 'necessity' does not imply reality; it is a weaker statement than truth. After all, some unknown celestial body may interfere with the moon's orbit in the meantime.

From the tense-logic thus outlined one can go on to form a theory of probability which confines the primary use of probability to a quantitative refinement of the modalities as applied to the future. This theory of course encounters the well-known difficulties that face any theory of empirical (observable) probabilities. A subjectivistic theory of probability describes what people may expect if they are thinking consistently. But that will not be enough for a physicist. He not only wishes to know whether the assumption of a certain probability for some event is logically consistent but whether it is correct. The question is what he means by calling the assigning of a probability value to an event correct. It ought to admit of an empirical confirmation, at least in principle. Probabilities are actually measured as relative frequencies in large series of observations under equal conditions. The main conceptual difficulty here is that the relative frequency itself, in any finite sample, is not strictly predicted by assigning a probability to the event. Thus its measured value is not supposed to be identical with the assigned probability, hence it is not a true empirical expression for it. However, this difficulty is not lethal for the theory. No theoretically predicted quantity can be compared directly with the result of one measurement of the same quantity. One must take account of the statistical distribution of measured values. In this sense any predicted quantity can only be verified empirically within some probability but not with certainty, the probability approaching the value one as the number of cases increases. We only have to admit that this is also the case for the particular quantity which we call probability. A probability can in this sense be defined as a prediction of a relative frequency, a definition which leads on to the more refined statement that it is the expectation value of a relative frequency. This expectation value is defined with respect to probabilities within statistical ensembles *of* statistical ensembles. This definition is not circular but has an 'open end' which is closed by assuming that some probability sufficiently close to one can practically be identified with certainty. I would be prepared in a rather lengthy discussion to defend the view that this description of probability goes precisely as far as one can go in a theory of experience.

3. *A set of axioms for a finitistic quantum theory is formulated, and discussed in connection with the preconditions of experience. 'Finitistic' here means having a finite-dimensional Hilbert space.*

In this section I make use of a doctoral thesis by M. Drieschner which is soon to be published.

Under the title of 'preconditions of any experience' I have so far collected the structure of time as described by tense-logic, and the concept of objective probability as applied to the future. It would be the ultimate goal of physico-philosophical ambition to proceed in the same manner towards further preconditions of all possible experience until their list would suffice to formulate the complete set of basic assumptions of a unified physics as postulated at the end of section 1. This would presuppose two achievements, neither of which I am able to present. One would be not a sketch, as given in section 2, but a full and thoroughgoing philosophical analysis of concepts like time, logic, number, probability. The other would be the extension of the analysis to some basic concepts of physics like object, state, change, observation. Such an analysis might lead towards a theory of the probabilities with which we can predict the changes of observable states of any object. The existing theory most closely corresponding to such a description is quantum theory. Thus it does not look *a priori* hopeless to deduce quantum theory from the preconditions of experience.

As long as this analysis has not been achieved, an easier heuristic method may be applied which uses the so-called axiomatic method as one of its main tools. One can try to find a set of axioms which on the one hand suffices to deduce the known system of quantum theory, while on the other hand admitting of a simple interpretation in terms of so-called preconditions of experience which, while they may not give the impression of being *strictly* necessary for any possible experience, still show some degree of plausibility. This is done in Drieschner's paper.

I should like to spend two additional preparatory paragraphs on the methodological meaning of axiomatics. It is a Greek discovery that mathematics can be organized in deductive systems, that is into sets of propositions which are divided into two classes: the axioms which are presented without a proof, and the theorems which can be logically deduced from the axioms. The tradition (with the notable exception of Plato) considered axioms to be evident and hence not in need of proof. This primitive attitude was psychologically shaken by the discovery in the nineteenth century that non-Euclidean geometry is logically possible. I say 'psychologically' shaken since the refuted hope of *logically* deducing the parallel postulate from the other axioms had in itself had its origin in a well-founded suspicion that it was not sufficient to state this particular axiom as simply evident. What was shaken was the trust in the simplistic notion of 'evident propositions'. Under the influence of Hilbert's 'Grundlagen der Geometrie' another, more sophisticated, but in my view still simplistic notion took its place and has been dominating the twentieth century. It is the view that it is no business of the mathematician to know whether a certain set of axioms is true or not, nor even what might be the meaning of calling it true or false. Certainly nobody can be forbidden to define terms such that the term 'mathematics' is restricted to the analysis of the logical connection between sets of formally presented 'propositions'. Yet under the attack of Brouwer, Hilbert had to admit that there is a field which is now called 'metamathematics' in which we study the presuppositions of 'mathematics'; these presuppositions are not 'axioms' but, e.g. 'ways of action' which we understand 'intuitively'. It seems that this field contains more than logic, e.g. the structure defined by the basic intuition of finite number. It is tempting to say 'true mathematics is metamathematics'; this would be the intuitionist standpoint. In any case, mathematics seems to face problems that are analogous to (and, as I personally think, basically identical with) those I am discussing here for physics.

In physics we encounter some theories, like quantum theory, in the shape of mathematical formalisms. Such a formalism admits of an axiomatic analysis. Yet, as pointed out at the end of my paper in Part II of the book, a mere formalism is not yet part of physics; it needs a physical semantics in order to be physics. If the formalism is given in axiomatic form it will suffice to give a semantic for the axioms. If this semantic refers to particular experience we will call it empirical. Our present attempt is to refer it not to particular experience but to traits common to all experience, perhaps even to 'all possible experience'. Yet in practice our attempt does not prove the necessity but only tends to show some plausibility of our axioms. In one respect (finitism, see postulate D below) we dare to rely more on our epistemological arguments than on the present form of quantum mechanics.

Drieschner's paper begins with a verbalized discussion of the preconditions of experience, then formulates a set of axioms in the language of mathematical logic, and ends up by deducing ordinary quantum mechanics (with one deviation) from the axioms. I shall condense the results of the first part into a group of 'postulates', after which I shall repeat Drieschner's 'axioms' in ordinary language, and comment upon their meaning and consequences. I will assign a symbol (letter or number) to each postulate or axiom, and in addition give it a name which loosely indicates its most important aspect. I will not now study the logical interlinkage of the postulates and axioms.

A. Postulate of Alternatives. Physics formulates probabilistic predictions for the outcome of future decisions within empirically decidable alternatives. To the extent that physics refers to probabilistic predictions for the future, this postulate states the result of our tenselogical considerations. That these probabilities intrinsically cannot be reduced to certainties is not postulated here; it will be explicitly stated in axiom 4c. The postulate adds that the probabilities refer to decisions within empirically decidable alternatives. An 'alternative' in the sense here used can be defined as a complete set of mutually exclusive temporal propositions. These propositions I shall also call the 'answers' of the alternative. They are temporal propositions in the sense of section 2. This means that in their simplest form they state something about the present ('it is raining in Hamburg'). They can be formulated so as to apply to some past, future or 'objectively defined' time ('it was raining in Hamburg yesterday', 'it will be raining in Hamburg tomorrow', 'it "is" ("was", "will be") raining in Hamburg on 1 July 1968'). Throughout our investigation time will be considered as admitting measurement by clocks which 'in principle' assign it a value at any moment, describable by a real variable t; the necessary scepticism against this simplification is beyond our present scope. Our temporal propositions and hence our alternatives are conceptually 'time-bridging' in the sense that it is meaningful to speak of deciding the same alternative ('is it raining in Hamburg') for different times. The words 'the same' evidently do not mean 'identical', since it is a different statement that it is raining in Hamburg the first or second day of July 1968. 'The same' means 'falling under the same concept' ('raining in Hamburg'). This definition of 'the same' will suffice only if we always take the concept as narrowly as possible (not 'raining', but 'raining in Hamburg'; not 'a particle is at position x' but 'this particle is at position x'); the use of demonstrative pronouns for defining a concept with sufficient narrowness may be inevitable. We consider alternatives as decidable without investigating how the decision is effectuated. It is only characterized as 'empirical', which means depending on particular ('contingent') phenomena that may occur or not occur at a given time. What may occur is considered formulated by the respective theory. Physics itself provides the sets of what I like to call 'formally possible temporal propositions' which formulate what may occur at all at any time. They are called 'formally possible' as distinct from 'actually possible' which refers to the future at some particular time; thus it is formally possible that it may rain in Hamburg any time, including this afternoon, but knowing the meteorological forecast I conclude that rain this afternoon in Hamburg is actually impossible. The answers of a given alternative must be mutually exclusive, i.e. if one of them is actually true all others must be actually false. We are only permitted to call a temporal proposition actually true or false for the time at which an alternative of which the proposition is an answer, is actually decided. It will be one of the main points of quantum theory that we are not permitted to assume the truth or falsehood of answers of sufficiently refined alternatives in situations where these alternatives are not actually decided. At present this remark is only a warning against reading an interpretation into the words 'true' and 'false' which is not to be implied; it will be excluded only by axiom 4c. I finally define that the answers to an alternative form a complete set if and only if the following holds: if all the answers of an alternative except one are actually false, that one is actually true. Our postulate entails that such complete sets exist, and we will confine the term 'alternative' to such sets. Thus terminologically an alternative is not confined to two possible answers. If the number of its answers is a positive integer n, we shall call it n-fold; a 2-fold alternative will also be called a simple alternative. For 'empirically decidable alternative' I shall briefly say 'alternative'. Our alternatives evidently correspond to what is usually called observables, and their answers to the possible values of the observables.†

† Much of the analysis of these concepts is due to Scheibe.[1]

B. Postulate of Objects. The answers of an alternative ascribe contingent properties to one object. Logically this means that the answers to our alternative can be formulated as 'categoric judgments', ascribing a predicate to a subject, here called object. The postulate is meant to entail that one alternative refers to one object, but that different alternatives may or may not refer to the same object. I think that in a sufficiently developed theory postulate B will be a consequence of postulate A, that means that the concept of an object will be reducible to the concept of (time-bridging) alternatives. This reduction is, however, not attempted in Drieschner's paper. Thus I here hypothetically accept 'object' as a basic concept of experience.

C. Postulate of ultimate propositions. For any object there exists ultimate propositions, and ultimate alternatives whose answers are ultimate propositions. Here an ultimate (contingent) proposition about an object is defined as a proposition which is not implied by another proposition about the same object (with the trivial exception of the 'identically false proposition' which by definition implies every proposition—*ex falso quodlibet*). The particular properties corresponding to ultimate propositions will be called 'states' of the object. For a classical object a point in phase space represents a state, and the set of all these points represents its (unique) ultimate alternative. For a quantum-theoretical object a one-dimensional subspace of Hilbert space represents a state and any complete orthonormal system represents an ultimate alternative. I here omit the very interesting consideration by which Drieschner tries to show that this postulate belongs to the defining properties of any physical theory.

D. Postulate of finitism. The number of answers to any alternative for a given object does not exceed a fixed positive integer n, characteristic of the object. This is the postulate in which Drieschner's approach differs from usual quantum theory. Of course one may expect that by choosing n sufficiently large it will be possible to avoid contradictions with the predictions of usual quantum theory. One may also expect that this postulate will lead to a great simplification of proofs in the ensuing axiomatic construction of the theory. But the two expectations might be mutually exclusive. If finitism leads to a true simplification of the theory it will probably exclude some features of possible objects of the theory which must be taken into account in an infinitistic theory and which are consequences of 'true' infinities. It is Drieschner's and my view that in such a case of conflicting consequences the finitistic view should tentatively be considered as the correct one. I think, the test case will only arise in the interaction problems of quantum field theory, for which compare section 4. At present I confine my argument for the truth of the postulate to the statement that no alternative with more than a finite number of answers can actually be decided by an experiment, and to the naive remark that a physicist ought to be surprised to find phenomena in nature in whose description the word 'infinite' would not admit of being replaced by the word 'very large'. For a more refined argument I refer to Drieschner's thesis, and, partly, to section 4.

E. Postulate of the composition of objects. Two objects define a composite object whose parts they are. The direct product of any two ultimate alternatives of the two parts is an ultimate alternative of the composite object. If a_k $(k = 1 \ldots m)$ and b_l $(l = 1 \ldots n)$ are the answers to the two alternatives, $a_k \wedge b_l$ ($\wedge$ means 'and') are the answers to their direct product. This postulate seems logically natural enough; it has far-reaching consequences in quantum theory.

F. Postulate of the probability function. Between any two states a and b of the same object a probability function p(a, b) is defined, giving the probability of finding b provided a is necessary. Language and content of this postulate rest on the more fundamental assumption that whatever can be said about an object in a manner admitting of empirical test, must be equivalent to the prediction of some probabilities. This is thought to be true because empirical test means precisely the decision of an alternative in the future, for which nothing but probabilities can be predicted. This view first leads to the phrasing 'provided a is necessary' for a condition usually expressed as 'provided the object is in the state a'. For the empirical test of the statement that the object is in state a is simply an experiment for which the outcome 'a' is predicted with certainty, that is $p(a) = 1$, which we identified in section 2 with 'a is necessary'. Besides this basic assumption the postulate expresses the view that the relationship between the states of the same object admits of an intrinsic description, that is a description which does not refer to any objects besides the one under consideration. One can also express that by saying: The states of an object remain always the same, independent of its surroundings.† That this must be true is perhaps not evident;

† Of course, 'states' here means 'Schrödinger states'; we say that the object can change its state with time, while the definition of the states through which it wanders can be given independently of time and—what matters here—of its surroundings.

it seems to be a rather strong statement on the meaning of the concept of an object. I shall accept it here as a principle. If it is true, the postulate becomes quite plausible, since what other intrinsic description, admitting of empirical test, should there be possible except the probability function?

G. Postulate of objectivity. If a certain object actually exists, one ultimate proposition about it is always necessary. The premise 'if a certain object actually exists' will turn out to be less trivial than it seems. 'A certain object' is to mean an object for which it is known how to decide its alternatives or at least some of them. Clearly if the so-characterized object does not exist at all, none of its states can be found and hence no ultimate proposition about it can be necessary. Now assume the object actually exists. This is a proposition about the object. We assume that every existing object admits of k-fold alternatives with $k > 1$. The proposition that it exists hence is not an ultimate proposition, for it is implied by any answer to one of its alternatives. We further assume that a theory about the object is possible which makes predictions about its behaviour; else the object would not fall under our concept of experience. The predictions can only consist in ascribing probabilities to the contingent propositions about the object. Any possible contingent theoretical statement on the object hence will be describable as a list of such probabilities; we call that a statistical characterization (s.c.) of the object. We now distinguish incomplete and complete statistical characterizations. In an incomplete s.c. some possible information is lacking; a complete s.c. gives as much information about the object as can be given according to its theory. Thus the knowledge that some state a is necessary is a complete s.c., which is expressed by the function $p(a, b)$ for all b. By definition $p(a, a) = 1$. Now consider any complete s.c. It is itself a contingent proposition on the object; let us call it c. It must be ultimate. Otherwise it would be implied by at least two ultimate propositions (if it were implied by precisely one, it would in turn imply that one and hence be equivalent with it); then an alternative might still be decided that goes beyond c, and c would not be complete. Now c implies itself ($p(c, c) = 1$), and hence one ultimate proposition is necessary. No two different ones cannot be necessary, since the s.c's corresponding to them would imply each other, reducing them to equivalence.

This rather delicate argument of Drieschner's turns on the distinction between complete and incomplete knowledge. In complete knowledge, probabilities different from 0 and 1 may (and will, according to axiom 4c) occur. They will *not* be described as being due to lack of information, while thermodynamical probabilities are thus described. This means that the idea of an 'objective situation' which can be either known or unknown is not given up in this theory, and therefore I have chosen the name 'postulate of objectivity'.

It must be seen, however, what consequences follow from the language here adopted. What 'propositions about the object X' are here admissible (X being a proper name of the object under consideration)? All ultimate propositions about X are admitted. Drieschner also shows that, according to his axioms, and using appropriate definitions of 'not', 'and' and 'or', a complemented lattice of propositions about the object can be built up. It can be described as consisting of all ultimate propositions and all finite 'logical sums' of the type 'a or b', 'a or b or c', etc. This lattice is a model of v. Neumann's quantum logic. 'Mixtures', however, are not permitted as 'propositions about the object' but only as 'propositions about a (real) collection of objects'; a point for which I again must refer to the thesis.

Another proposition that cannot be accepted as a 'proposition about X' is any ultimate proposition about a composite object consisting of X and, say, Y, that is not (speaking in terms of completed quantum theory) a product of pure states of X and Y. This must be so, because in a state of the composite object that is not a product state, none of the ultimate propositions about X is necessary. The solution of this apparent paradox (which is essentially the Einstein–Rosen–Podolsky paradox) is that in such a state it is simply not permissible to say that object X actually exists. It only has 'virtual existence', i.e. the composite object might be divided into X and Y, but only by altering its state. This is the relevant sense of the premise of postulate G. One cannot then say that X 'does not exist at all', but it does not 'exist actually'. I admit that I am not yet fully satisfied about the stringency of these distinctions, but I have so far not been able to do better.

I turn to Drieschner's axioms.

1. *Axiom of Equivalence. If a and b are ultimate propositions, $p(a, b) = 1$ is equivalent to $a = b$.* This is a two-way implication. '$a = b$ implies $p(a, b) = 1$' means $p(a, a) = 1$, as formerly stated. On the other hand, '$p(a, b) = 1$' means certainty that b will be found (in an experiment appropriate to finding b) if a is necessary, and this

is what we mean by saying that a implies b. But if b is ultimate, it cannot be implied by *another* proposition, hence $a = b$. We see in this simple example how these 'axioms' tend to express certain consequences of the basic assumptions formulated in the postulates.

2. *Axiom of finite Alternatives. If n mutually exclusive ultimate propositions a_i ($i = 1 \ldots n$) are given, then for any ultimate proposition b:*

$$\sum_{i=1}^{n} p(b, a) = 1.$$

This is a probabilistic version of what was meant in the commentary on postulate A by attributing a *complete* set of mutually exclusive answers to any alternative, combined with postulate D. For two mutually exclusive ultimate propositions a and b we evidently must assume $p(a, b) = p(b, a) = 0$.

We now introduce 'propositions about X' as *sets* of 'ultimate propositions about X' (for the sake of brevity we will from now on omit 'about X'). A proposition A then means 'the necessary ultimate proposition x is an element of the set A'. Here postulate G is used in speaking of 'the necessary ultimate proposition'. $\bar{A}$ (i.e. 'not A') is defined as the set consisting of all those ultimate propositions which are mutually exclusive with all elements of A. $A \wedge B$ ('A and B') is the set of all elements *contained* both in A and in B. $A \vee B$ ('A or B') is defined as $\overline{\bar{A} \wedge \bar{B}}$.

3. *Axiom of Decision. For any A there exists an alternative $a_1 \ldots a_n$ such that $a_1 \ldots a_e$ are elements of A while $a_{e+1} \ldots a_n$ are elements of $\bar{A}$.* We have assumed in postulate A that all contingent propositions are decidable. We now assume that there are always ultimate alternatives 'adapted' to any decision between A and $\bar{A}$. Here it must first be seen that A and $\bar{A}$ form a simple alternative. This is not trivial since most of the ultimate propositions will belong neither to A nor $\bar{A}$. But if A is decidable, an experiment is possible by which either 'A' or 'not A', whatever 'not A' may turn out to mean, is found. If 'A' is found, then for an immediate repetition of the experiment A will be necessary; if 'not A' is found, A will then be impossible. This is what we mean by saying that the statement 'A is decided' is subject to empirical test. Now 'A' means that the necessary state belongs to the set A. 'Not A' then must mean that it belongs to a set of states excluding all states of A. The decidability of A also means 'not not

$A = A'$, hence the states of A also exclude those of 'not A', and 'not A' must indeed consist of all states mutually exclusive with those of A, that is 'not A' $= \bar{A}$. I omit the more complicated argument leading to axiom 3 in its fullest sense.

4a. First Axiom of Completeness. For any set of $k < n$ mutually exclusive ultimate propositions $a_1 \dots a_k$ there exists an ultimate proposition a with $p(a, a_i) = 0$ $(i = 1 \dots k)$. This axiom together with axiom 2 implies that all ultimate alternatives have exactly n answers. It may also be called an axiom of symmetry for ultimate alternatives.

4b. Second Axiom of Completeness. For any set of $n-2$ mutually exclusive ultimate propositions $a_3 \dots a_n$ and any ultimate proposition b there exists an ultimate proposition a_2 which excludes all a_i $(i = 3 \dots n)$ and b, i.e. $p(a_2, a_i) = 0$ and $p(a_2, b) = 0$. These two axioms seem rather special. Perhaps they are not very strong statements since certainly no ultimate alternative with more than n answers is acceptable, and alternatives with less than n answers can always formally be completed by adding some ultimate propositions which, due to the assumption that the given alternative has less than n answers, must have probability 0 for all possible measurements. However, axiom $4c$ will in its general formulation exclude these cases, and it will in this sense be the truly strong statement.

4c. Axiom of Indeterminacy. For any two mutually exclusive ultimate propositions a_1 and a_2 there is an ultimate proposition b such that $p(b, a_1) \neq 0$ and $p(b, a_2) \neq 0$. The weaker statement 'there is a pair of mutually exclusive a_1 and a_2 such that...' would suffice to introduce indeterminacy, i.e. to exclude the trivial classical model of the other axioms in which all ultimate propositions are mutually exclusive. I feel that indeterminacy is a precise expression of what I have called the openness of the future in sections 1 and 2. Without this assumption the application of probability to the future would reduce to lack of knowledge, and I would not see how the second law of thermodynamics might then be reconciled with our assumptions. A reader who is not convinced by this argument may have to accept the axiom pragmatically because it leads to quantum theory. Its general formulation 'for any a_1 and a_2...' introduces full symmetry in the space of states. Concerning this symmetry cf. section 4, postulate M.

5. *Axiom of exclusion. For ultimate propositions, $p(x, y) = 0$ implies $p(y, x) = 0$.* This is connected with the law of double negation $\bar{\bar{A}} = A$.

Using these axioms Drieschner can show:

(*a*) The set of propositions is a complemented non-Boolean lattice.

(*b*) It is a projective geometry of $n-1$ dimensions.

(*c*) It is isomorphic with the lattice of the subspaces of an n-dimensional vector space.

(*d*) In this space the probabilities define a metric.

We are then left with the question on what algebraic number field this vector space is erected. I formulate the basic assumption as

H. Postulate of measurability. Any linear operator in the vector space that leaves the metric invariant is an observable and

I. Postulate of continuity. The possible changes of state in time are described by one-to-one mappings, continuous in time, of the set of states on itself.

The probability metric together with postulate I show that the number field must contain the real numbers, leaving only a choice of the real numbers themselves, complex numbers and quaternions. If we try to describe measurements in our theory we are led to conclude from postulate H that any admitted linear operator must permit diagonalization. Hence the number field must be algebraically closed, which leads to complex numbers. The result of these considerations is formulated by Drieschner in two axioms which I do not repeat here.

One may wonder how well postulates H and I are founded in the preconditions of experience. I cannot here explicitly argue for I, which contains both continuity and the basis for reversibility. This would entail a return to the theory of time on a higher level than in section 2, which I have not achieved so far. Postulate H repeats v. Neumann's famous general identification of operators and observables. In infinite-dimensional Hilbert space this was a sweeping assumption that was apt to provoke scepticism. In our finite-dimensional Hilbert spaces it might look more plausible. Since certainly *some* operators will be observables, it again expresses some high symmetry of the manifold of states. The question why we assume symmetry must be asked once more under a new heading.

4. *Finitistic quantum theory is interpreted as directly implying a unified physics in which cosmology and elementary particle physics are necessarily connected.*

The three preceding sections presented an analysis of existing physics. Now I offer an hypothesis on the path that will lead beyond it. We are seeking a unified theory that would produce the concepts and laws of the five theories discussed in section 1 out of a few universal principles. The hypothesis says that the conceptual tools for this task are already assembled in our former analysis.

I assume that the preceding analysis has sufficiently clarified quantum theory (no. 2 in section 1), thermodynamics (4), and their mutual relationship, essentially reducing them to theories on the deciding of alternatives in time. 'Sufficiently' does not mean 'completely' since that, if possible at all, would presuppose a full philosophy of time, but it means sufficiently for proceeding to the three remaining theories. We still have to introduce space both locally (1) and at large (5), and ultimate objects (3). I think this can be done by a consistent use of the one postulate in which our presentation differs from usual quantum theory, i.e. the postulate of finitism (section 3, D).

(2 and 5): I shall begin with the question of the totality of objects and of space at large, that is of cosmology, stating a *preliminary* postulate:

J. Postulate of approximately stationary cosmology. The universe can be approximately treated as an object. Treating the universe as an object is exactly what cosmological models do. The postulate will serve a similar purpose in the present theory: it will permit an easy way of expression. Yet the identification of the universe with an all-embracing object cannot be exactly correct, and in order to avoid misunderstandings I shall formulate the objections against it before using it (cf. the corresponding passage in section 1).

An object is an object for subjects in a world. This would be made even clearer if the concept of an object could be reduced to decidable alternatives (see commentary to postulate B). It seems philosophically absurd to reduce the world to what is possible only within a world. A more particular objection is that our concept of an object only contains 'time-bridging alternatives'. Hence no contingent questions can be asked about a given object in the future that would not have been equally possible in the past; further on, the postulate of finitism treats the number of answers of admissible alternatives as constant in time. As an example, let us briefly consider the meaning of continuity in a finitistic theory. It is certain that no alternative that has ever been decided or will ever be decided will have more than a finite number of answers. But this number is not *a priori* fixed. A continuum, say a line, may have actually been divided in the past into no more than n parts, thus e.g. answering the n-fold alternative 'in what part of the line is the material point P?' (It is to be remarked that I always speak of a *physical* continuum, hypothetically supposing that physical continua exist.) Yet continuity implies that it is possible to go beyond any fixed number n of the parts; we can go on dividing the line. The words 'it is possible to...', 'we can go on' refer to the future. The next division will again be finite, but its n' may be larger than n. In this sense a continuum is not an object, if we define objects according to section 3; it is rather an indefinite number of possible objects. The basic idea of the postulate of finitism can then be formulated as the converse statement: a given object is not continuous. That one should treat given objects as non-continuous is, I think, the fundamental truth of quantum theory. Quantum theory has eliminated the trouble created by continuity in classical physics (Planck), and it has run into trouble itself where it attempted to build 'true' continuity into its frame, that is in the field theory of interactions. (A non-relativistic theory of mass-points in continuous space does not describe 'true' continuity since it would not be essentially changed by replacing space by a point-lattice.) The word 'given' in the phrase 'given objects' means what is already given, that is it refers to the past as assembled in the present. Finitism says that only a finite number of answers has been actually given at any time. It should not rule out that the number of possible contingent propositions is infinite in the potential sense: for any given number of propositions another proposition can be formulated and empirically proved. In this sense continuity means essentially the openness of the future. But then in a continuum, the number of 'objects' cannot be constant in time. Since we do not wish *a priori* to rule out continuity as a trait of the real world we should not assume the world to be strictly an object.

(2, 3, and 5): Given these precautions we now draw some conclusions from postulate J, neglecting the word 'approximately' for the moment. Let the universe be an object with an n_u-dimensional Hilbert space. Our first question is how we can at all introduce a concept like position space or configuration space into the theory of such an object. I shall first discuss this question in a formal manner.

Historically, quantum theory was constructed the other way. Configuration space was thought of as given, and vectors in Hilbert space were an abstract description of functions in configuration space. We now successively ask what conditions must be imposed on configuration space in order that the number of dimensions of the Hilbert space of its functions should be (a) denumerable, (b) finite. If I am not mathematically mistaken, condition (a) means that the configuration space must have a finite volume. Uusually one imposes a periodicity condition. It would seem more natural to use a closed configuration space; this will be my working hypothesis. Then condition (b) can be fulfilled by a cut-off, e.g. by using only the n_u first functions of some orthonormal basis. Assumption (a) is a cosmological assumption, assumption (b) looks like elementary particle physics. It is my hypothesis that this is not only apparently so. I think that finitistic quantum theory implies (a) a cosmology, (b) ultimate objects, entailing some sort of an 'elementary length' and elementary particles, (c) a necessary connection between both theories.

(1 and 2): What is the physical semantics of space? Configuration space is just a mathematical description of the set of positions of mass-points in space. What is the meaning of 'space'? I think it is the basic concept of the theory of a manifold of interacting objects. Inverting postulate E we say that an object can be considered to consist of objects which we call its parts. The inner dynamics of the given object must then be described partly as the inner dynamics of each of its parts, partly as their interaction. Let us for simplicity speak of two objects A and B which are the two parts of an object C. Let A have an m-dimensional Hilbert space, B an n-dimensional one, then C's Hilbert space will have $m \cdot n$ dimensions (I so far neglect problems like Fermi- and Bose-statistics). The conceptual splitting of C into A and B has the practical meaning that we do not always need to decide mn-fold alternatives but that there may be practicable measurements of m-fold alternatives of C which permit the expression that they are 'alternatives of A'. Practically this means that the interaction of A and B must be small enough while a measurement on A takes place. We can use the approximation of isolated or 'free' objects while on the other hand we know that no absolutely isolated object can be an object for us (known to us). Thus the concept of interaction corresponds to the approximative way of talking on which all physics rests: we speak of separate objects or separate alternatives, knowing that they do not exist in a strict sense, and we correct this mistake by describing them as interacting objects.

I now maintain that the continuous parameter on which the interaction depends is called position, and that the manifold of the possible values of this parameter is the meaning of 'space'. I shall rely on concepts belonging to the language of particles for an illustration. Let us consider the composite object C in the absence of additional, 'external' objects. In order to speak meaningfully of A and B as parts of C, we need to posit 'separated' states of C, that is states in which the interaction between A and B is negligible. In order to speak meaningfully of C as consisting of A and B, we need to posit 'connected' states of C, in which the interaction between A and B is not negligible. Let us assume that C goes through a 'scattering process', i.e. for $t \to -\infty$ and for $t \to +\infty$ the states shall be separated, but in some time interval from t_1 to t_2 they shall be connected. How does the change of state of C determine the value of t_1, that is the time of the beginning of interaction? This presupposes (a) that the states of A and B will change even without interaction. This I call the weak law of inertia; it seems to be a presupposition of the concept of separate objects. Further it is presupposed (b) that t_1 is defined by values of changing parameters that are defined even in the separated states, that is of parameters defined for 'free motion'. Let there be such a parameter x for A and y for B, then the value of the interaction energy must depend on some relation between x and y. I maintain that position will turn out to be this parameter. That means: there is not, first a parameter called position, and second, an interaction that happens to depend on position; but rather: there is a parameter on which interaction depends, and that is what we mean by position. Of course this must be shown in mathematical detail. Here a third presupposition enters: (c) if the theory of free objects is invariant with respect to some group, then the interaction must be invariant with respect to the same group. This must be so because the relation between x and y on which the interaction depends must be invariant with respect to the free group, since x and y are defined for free objects.

This consideration is only heuristic, for it depends on the assumption that there are 'separated' and 'connected' states, which will not be true for ultimate objects. I have not been able to replace it by a stricter one. In any case it indicates that we will have to introduce space and the plurality of objects together.

(1, 2, and 3): How far can we go in splitting the objects? We try to assume the more radical hypothesis:

K. Postulate of ultimate objects. All objects consist of ultimate objects with $n = 2$. In German I called them 'Urobjekte' and their alternatives 'Uralternativen'; as a whimsy, I proposed the abbreviation 'ein Ur' = 'an ur' for such an object, which I shall use here too. This postulate is trivial as long as we do not specify the law of interaction for ultimate objects. A Hilbert space of n-dimensions can always be described as a subspace of the Kronecker product of at least r 2-dimensional Hilbert spaces, where $2^{r-1} < n \leq 2^r$; the $2^r - n$ unused dimensions can be excluded by imposing a law of interaction that admits a super-selection rule between the two subspaces. The ultimate objects become meaningful if we use our previous heuristic consideration for stating the

L. Postulate of interaction. The theory of the interaction of ultimate objects is invariant under the same group as the theory of free ultimate objects. This is a strong, non-trivial assumption. It first obliges us to study free urs.

A single ur is an object with a two-dimensional Hilbert space. It admits the transformation group SU_2. I should like to emphasize the physical meaning of this primitive mathematical statement. The 'theory of a free ur' describes not just its manifold of states but also its law of motion. The equation of motion must be invariant under SU_2. It is easy to find the solutions of such an equation; the state vector can only assume a common time factor $e^{-i\omega t}$ with state-independent ω, hence the states themselves, being one-dimensional subspaces, remain unchanged in time. Yet the question is how we know this to be the correct condition to impose on the law of motion. This is the form in which we now face the question of section 3, why we should postulate symmetry of the state-space. I formulate the principle here used as a postulate:

M. Postulate of symmetry. For a single ultimate object none of its states is objectively distinguished from the other. An 'objective' distinction here is a distinction 'by law of nature' as opposed to a 'contingent' one like e.g. 'the state in which this object is just now'; hence the *law* of motion cannot distinguish the states. What is meant here also permits the following description: an ultimate object is defined by a simple ultimate alternative. To distinguish between two of its states means to define at least one additional alternative. If an answer of this additional alternative implies one of the states of the original ur, the proposition that this state obtains is implied by another proposition. Hence it was not ultimate. Hence not a single ur was present but at least two of them. In this sense the possibility of speaking of a separate object simply means that its state-space is symmetrical; else it is not a separate object. To put it still differently: the approximation in which we can describe objects as symmetrical is the same approximation in which we can describe them as separate. Physics is precisely the attempt to describe reality starting from this approximation. Perhaps this approximation is nothing more than the precondition for the use of concepts.

The Hilbert space of a single ur is a two-dimensional representation space of the SU_2. Postulate L implies that the Hilbert space of many urs must be a higher-dimensional representation space of the same group. The law of motion of interacting urs must be invariant under the simultaneous transformation of all urs by SU_2. Hence we seek a mathematical description of the states of urs which from the outset will express this invariance. We get it from the so-called regular representation of the group. In this representation the elements of the group are linear operators acting on a vector space whose vectors are linear combinations of the group elements themselves. Thus the 'regular' representation of state vectors of urs will describe them as functions on the SU_2. Now this group is a real three-dimensional spherical space. Only invariants of relative position in this space can serve as parameters on which interaction of urs can depend. Hence it must be identified with position space.

Thus our postulates force us to assume that position space (that is what we usually call 'space' or 'cosmic space') is a three-dimensional real spherical space. I take this to be the reason for the three-dimensionality of space; a quantum theory of ultimate objects with $n = 2$ would not admit of another invariant description.

We will now have to draw the consequences of our postulates for the three theories 1, 3, and 5.

(3 and 5): 'Elementary particles' must be built up from ultimate objects. The urs would be something like elements of a Heisenbergian 'Urfeld'. My attempts in this field have so far not been successful enough to justify a description. I should merely say that this elementary particle theory is cosmological from the outset. The curvature of space means a quantization of momentum. The single ur corresponds to a particle with minimum momentum; it is hence not localized in cosmic space. If we assume a radius of the world $R = 10^{40}$ nuclear units of length $= 10^{40} \cdot 10^{-13}$ cm, a particle of

nuclear momentum or a particle localizable in a nucleus would have to consist of about 10^{40} urs. The total number of urs in the universe might be tentatively identified with the number of bits of information possible in the universe. Assuming for the sake of simplicity that there is just one sort of elementary particle, say a nucleon with Fermi-statistics, one might estimate that there are as many bits in the world as there are cells of nuclear size, each of which can either be occupied or empty. This number is $N = R^3 - 10^{120}$. The number of dimensions of the Hilbert space of the universe then would be $n_u = 2^N$. If one nucleon consists of R urs, there ought to be $R^2 = 10^{80}$ nucleons in the world. I do not yet dare to take the good agreement of this figure with experience as a confirmation of an hypothesis which is not sufficiently elaborate.

The cut-off used in limiting the total number of urs is not simply interpreted as an 'elementary length'. The precision with which we can determine a length depends on the momentum involved in the measurement, that means in this theory on the number of urs present in it. If we wish to localize all nucleons of the world simultaneously, each one is assigned 10^{40} urs and hence occupies 10^{-13} cm. If we wish to measure a length more accurately we have to apply momenta corresponding to an energy higher than the rest energy of the nucleon. Assuming that all urs of the world are used to measure one length, this might lead to an accuracy of 10^{-93} cm. In no case, however, will this theory describe space as a discrete lattice.

(1, 3, and 5): Elementary particle physics essentially depends on its symmetry groups. It would have to be the goal of the present theory to deduce these from postulate M. That would presuppose a model of how particles are built up of urs; such a model I do not offer here. But there seems to be a difficulty from the outset. The most generally accepted group in field physics is the Lorentz group; and the present theory is not Lorentz-invariant. Looking at its basic assumptions, Lorentz invariance was indeed not to be expected. We started with an absolute concept of time as one does in non-relativistic quantum theory. We have arrived at a cosmology with a static universe, and such a cosmological model in its turn is again not Lorentz-invariant. But it should not be difficult to find a law of interaction obeying postulate L which would make the theory locally Lorentz-invariant. One can take the view that this is all we should ask for. On the other hand our theory will not be semantically consistent without a description of the measurement of time. It is possible that this consideration might induce us to change our basic assumptions such that the resulting cosmology would permit a group which contains the homogeneous Lorentz group, e.g. the de Sitter group. This question, which I leave unsolved here, leads us back to the approximate character of postulate J. I shall end by formulating a last hypothesis:

N. Postulate of expansion. In second approximation the universe can be described as consisting of ultimate objects whose number increases with time. This formula takes account of the objections I raised in commenting on postulate J. It expresses the open future: if we wait long enough, we will be able to divide up a line very finely. It also takes account of certain cosmological objections.

(1 and 5): It would seem strange for a theory built on principles of very high generality to contain a basic constant whose value is contingent: the number N of the urs in the universe, $N = 10^{120}$. According to the new postulate, N would now be a measure of the age of the universe. If we measure this age t by ordinary clocks in nuclear time units, probably $N = t^3$. This formula is derived from the answer to a second objection: according to postulate J, the radius of the world should be a constant while astronomical evidence would favour the equation $R = t$. In fact, in the present theory, expansion of the universe is necessarily connected with the creation of matter by the formula $N = R^3$. The theory thus resembles the cosmology proposed by Dirac and Jordan.

A final word must be said on gravitation. The theory considers the topology of space at large as a consequence of quantum theory. To be more precise it considers the finitude of space as a natural expression of the finitude of experience. On the other hand the empirical connection between the curvature of space and the cosmic density of matter is sufficiently close to that demanded by Einstein's equation to seem more than accidental. As a possible explanation I mention the idea that it is not the curvature of space which adapts to a given density of matter with a given constant of gravitation, but that this so-called constant adapts to a given density of matter and curvature of space; Eddington's famous relation, expressed in our theory, says that the constant of gravitation, measured in nuclear units, is $g = R^{-1}$. A time-dependence of g has been discussed by Jordan; and R. Sexl has pointed out to me that in some theories gravitation is a many-body force, such that the constant g will appear to depend on the present density of matter.

In the present theory one may expect the following theory of gravitation. Elementary particle physics will lead to fields with many different transformation properties; among these there will also be a field behaving like the gravitational field (probably containing scalar as well as tensor components). According to Thirring (see section 1) a semantically consistent theory of this field will couple it with the metric tensor. This equivalence of gravitation and metric implies that our original assumption of a curved cosmic space merely says that there is a universal gravitational field. Einstein's theory of gravitation still contains one apparently contingent element: the value of the gravitational constant. The appearance of this constant is due to a dichotomy which Einstein never desired to retain but was unable to overcome: the dichotomy between matter and the metric field. He had to describe their mutual dependence as some sort of interaction. The influence of the metric field on matter he was able to describe in a way that avoided contingent constants by demanding the motion of matter to be along geodesics. The influence of matter on the metric field, however, took the form of a Poisson equation with a non-dimensionless constant factor. One would expect that in a semantically consistent theory this constant would reduce to a dimensionless factor whose nature is determined by the theory. If we measure all quantities referring to matter in nuclear units, g empirically turns out to be R^{-1}. In the present theory this is precisely what we would expect, once the identity of gravitation and metric has been accepted. As long as we neglect local gravitational fields and confine ourselves to the cosmology presented in this section, the relation follows directly from the assessment of N, that is of the amount of information present in the universe. In the language used here, Einstein's equation as applied to cosmology simply *means* that there is coupling between the amount of matter and curvature of space ($N = R^3$), due to the fact that the total volume of cosmic space, measured in nuclear units, *is* the number of bits in the world, and that ultimate objects are simply bits. Local gravitational fields are not yet contained in this description since they must be linked with the general theory of 'elementary' fields. Accepting Thirring's gauge transformation, however, we may expect that a local gravitational field signifies that matter locally behaves as if it were part of a universe with a metric expressed by the local values of the metrical field. In such a universe the density of matter would be different from the cosmic average, and precisely this connection between the mass tensor and the metric field is expressed by Einstein's equation.

REFERENCE

(1) Scheibe, E. Die kontingenten Aussagen in der Physik, *Axiomatische Untersuchungen zur Ontologie der klassischen Physik und der Quantentheorie* (Frankfurt am Main, 1964).

A PHILOSOPHICAL OBSTACLE TO THE RISE OF NEW THEORIES IN MICROPHYSICS

MARIO BUNGE

Everyone seems to agree that new theories, even more powerful than the present set of quantum theories (henceforth QT), are needed, since nature has proved to be far richer than it was thought four decades ago. The big question is: what kind of new theories do we want? This, in turn, poses a number of strategy problems, some of which are philosophical questions and must therefore be handled with philosophical tools.

Thus, one will start by asking whether we really need new general theories to replace QT or, rather, new—deeper and more sophisticated—models of the entities concerned (e.g. kaons), models that can be embedded in those general frameworks, thus providing as many theoretical models (specific theories) as needed. Should this move fail because of incurable defects (e.g. divergences) in the existing general theories, the next question arises: What should the future fundamental theories look like? May we keep the basic traits of QT or should we perform a thorough shake-up of the foundations? In particular, can we retain the basic stochastic character of QT or is it worth while looking for scatter-free basic variables? And can we borrow the spacetime of macrophysics, as we have done heretofore, or should we explore microphysical spacetimes? Shall we continue to theorize about unobservable entities, such as electrons and neutrinos, and unobservable properties, such as position densities and field variables, or shall we concentrate on observable entities and on directly measurable variables? Finally, can we afford to have thoroughly physical theories—i.e. theories about physical systems only—or is it indispensable to reckon with a subjective component? In other words: can microphysical theories concern independently existing things—as classical physics does—or do they have to concern the observer as well?

Of all these strategy problems the latter is the only strictly philosophical one. Since this problem—the one of the objectivity of physical theory—is usually handled in a careless way, I shall tackle it here using some results of philosophical semantics and the foundations of physics. A fuller treatment is given elsewhere.[1]

1. What is the referent? Once upon a time physicists were so unsophisticated that there was no doubt in their minds concerning the object of their inquiries: it was, they claimed, physical reality, and they strove to capture the invariant patterns, or laws, of that basic stratum of reality. They now know better: almost anyone will tell us that physics is *not* about physical reality—a dangerous metaphysical monster. Unfortunately, we are not certain what physics *is* about. Some authorities such as Heisenberg† maintain that QT is not about nature but about our knowledge of nature—what philosophers call epistemology. Others, such as Bohr,‡ say that physics is neither about nature nor about human knowledge but about a *tertium quid*, namely the 'quantum phenomenon', which would be an unanalysable (hence mysterious) unit somehow made up of things, pieces of apparatus, and observers. Still others believe that QT is about propositions, or about languages, or what not.

As a philosopher I am not bewildered by this variety of opinions concerning the subject matter of QT, for we are used to such a cacophony in philosophy. What I do find strange is that none of the authoritative—but, alas, mutually conflicting—opinions on such a grave matter is supported by sound *argument*: that all those *doxa* are held merely on the word of their propounders. This reliance on authority is at variance with the scientific attitude and it is utterly unphilosophical. Let me therefore handle in a philosophical way the question: *what is QT about?*

To a philosopher, this question calls for a semantic analysis of typical expressions of the language of QT. Such an analysis should result in disclosing both the connotation (content, intension) and the denotation (scope, extension) of the idea concerned—i.e. of the concept or the statement that the given expression designates. While the connotation of an idea is determined by the assumptions that characterize it, its denotation consists of the set of entities the idea fits. This latter set is called the *reference class* of the idea in question, and an

---

† E.g. Heisenberg[2] 'The laws of nature which we formulate mathematically in quantum theory deal no longer with the particles themselves but with our knowledge of the elementary particles.' Also Heisenberg,[3] for the need to give up 'the reality concept of classical physics'.

‡ Bohr remarks[4] that the epistemological lesson of atomic physics consists in that it shows that it is no longer possible 'to speak of an autonomous behaviour of the physical object'. Bohr may also be consulted[5] for the 'essential wholeness of a proper quantum phenomenon'.

element of this set is called a *referent* of that idea. Thus the reference class of continuum mechanics is the set of all bodies, while the one of electrodynamics is the set of all triples constituted by a pair of charged bodies and the associated electromagnetic field. Our original question can then be reworded as follows: *what is the reference class of QT?*

A customary way to answer this question is to collect authoritative statements found in popularization ('philosophical') writings (for a useful collection see reference (6)). This method has the advantage that it precludes any possibility of mistake—assuming that authority makes the truth. The philosophical method, on the other hand, is risky, for anyone can criticize its results. This other method consists in analysing the ideas (e.g. formulas) in question (conceptual analysis) and in watching how they fare in the laboratory (empirical analysis). While the former analysis should tell us what the *intended* or hypothesised reference class is, the second analysis will tell us how well represented reality is in this class, i.e. what the *actual* extension of the idea is. In many cases the actual extension of an attractive idea proves to be empty.

Take, for instance, the Hamiltonian H and the state vector Ψ in ordinary QT. Since those two 'quantities' are related by the Schrödinger equation, they must have the same reference class— roughly, they must depend on the same dynamical variables. Now the reference class of H, hence of Ψ as well, will depend on the problem at hand. Here go some standard examples in non-relativistic QT.

(i) *Free particle of the kind P* (e.g. proton)

$$R(\Psi) = P$$

where R stands for the reference function (a typical concept of semantics) and P for the set of all (actual and possible) particles or, rather, pseudoparticles, of the given kind.

(ii) *Particle of the kind P in a potential well* (e.g. alpha particle in a radioactive nucleus). Since the well represents in a global (phenomenological) way an environment of a certain kind E, acting on the members of the class P, we have

$$R(\Psi) = P \times E$$

where $P \times E$ is the Cartesian product of P by E, i.e. the set of all ordered pairs ⟨particle of the kind P, environment of the kind E⟩.

(iii) *Particle in an external field* (e.g. the Stern–Gerlach experiment). Here too a pair of physical systems is in question, but now the milieu is a specific one, namely a field of a given kind F, so that

$$R(\Psi) = P \times F.$$

(iv) *Scattering of a particle by a target.* The same as (iii).

(v) *Compton scattering.* The same as (iii), where now P is the set of electrons and F the set of electromagnetic radiation fields.

(vi) *Nucleon–nucleon interaction.* In the older theories (either phenomenological or involving specific nuclear forces) H was about a two-nucleon system, so that

$$R(\Psi) = N \times N$$

where N is the set of all nucleons. In the meson theory of nuclear forces we have to do with triples nucleon-pion-nucleon, so that here

$$R(\Psi) = N \times \Pi \times N$$

where Π is the class of pions.

In every one of the above typical cases the referent has turned out to be an individual physical entity, or else a couple, or a triple, of physical objects. In quantum statistical mechanics the reference class will be a set of N-tuples of individuals of a certain kind P, where N is a (fixed or arbitrary) positive integer. That is to say, here $R(\Psi) = P^N$. Finally, in relativistic theories we must count the reference frame among the referents of most 'quantities'. Thus instead of (i) we shall have $R(\Psi) = P \times K$, where K is the set of physical reference frames. Similarly with the other cases.

The result so far is disappointingly simple: *the referents of QT turn out to be n-tuples of physical systems.* No trace of an observer's mind. I might, of course, proclaim from the peak of philosophical wisdom that the referent of every quantum-theoretical expression is the sealed unit constituted by the microsystem of my dreams (an extremely tame one), my mother-in-law, and my own immortal soul. But people would start asking: how do you know? And I would be at a loss to cite any authority.

2. How is the referent recognized?

How does one recognize the referent of a given H (or Ψ)? If one is haughty and does not accept the gift of an authoritative answer, one will start by looking at the independent variable(s) and at the indexing variable(s) that label or name the referents. The least one will find is a single referent—e.g. a free photon; the most one finds is an actual ensemble of entities of various kinds—e.g. electrons wandering inside a crystal lattice.

However, the procedure of looking at the explicit independent variable(s) and index set(s) is not fool-proof, for very often some of the referents are assumed to make no important contribution to the Hamiltonian, so that their coordinates are ignorable. For example, in a semiclassical treatment, the nucleons in an atomic nucleus may be regarded as having fixed positions, so that only the intermediary meson will contribute to the energy changes. But, of course, when drawing the Feynman diagram of the interaction one is not supposed to forget the straight lines representing the nucleons. (Also, one is not supposed to take seriously every line in such a graph: some of the lines have no real referents.)

In general, whenever a variable of a system occurs in a Hamiltonian, the latter refers to that system. When no such variable occurs in H, one looks for it in the context, i.e. among the assumptions made when posing the problem. The prime rule for finding the possible referent(s) of a physical 'quantity' or of a physical formula is, then, this: *analyse the given expression in its context*. (Hence, do not look elsewhere, e.g. into 'philosophical' digressions.) That is, the theory as a whole will point, in a more or less explicit way—depending on the explicitness of its formulation—to its own *intended* referent. But this yields only a possibility: this possibility must be checked.

A first check is a purely theoretical one: one must make sure that the hypothesized independent variables, corresponding to the intended referent(s), are not idle. That is, one must see whether they make a difference to the results (e.g. the energy levels). This holds, in particular, for the 'variables' peculiar to the orthodox interpretations of QT. Thus if someone holds that a given H represents the energy measurements performed on an atom, one must challenge him to point out the apparatus coordinates occurring in H.

The second check of the hypothesis that the given formula has such and such a referent is, of course, of an empirical nature: the performance of the theory in predicting states and events will tell us to what extent the intended reference class of the theory overlaps with its actual reference class, i.e. with the extension or domain of validity of the theory.

In short, finding the referent of any piece of QT is a complex operation, every step of which is open to criticism. To assign referents by decree, without regard to the mathematical structure of the construct concerned, or to the assumptions that surround it, or to the relevant empirical tests (if any), is an arbitrary procedure that only very, very famous scientists can afford. The masses have got to find the referent by hard toil.

3. Where do the apparatus and the observer come in? What about the apparatus and the observer? Have we not been told that they are among the ever-present referents of every quantum-theoretical formula, to such an extent that it makes no sense to speak, say, of an atom except when subjected to observation? True, but a semantical analysis of the dynamical variables and state vectors of QT fails to uphold this tenet. Indeed,

(i) In most of the situations treated in QT no apparatus is involved. Recall the first two and the last two examples in section 1.

(ii) When an apparatus is in fact involved (referred to by a formula), then it is treated as one more physical system, satisfying physical (not psychological) laws, and distinguishable from both the main object (e.g. the scattered beam) and from the observer in charge of the experimental set-up. Recall examples (iii) and (iv) in section 1. By the way, these two examples fit also spontaneous processes, i.e. processes not under experimental control—which disproves the claim that in QT there is a chasm between natural processes (von Neumann's type II ones) and artificial processes (type I ones), a chasm that would be due to the action of 'the observer'.

It might be retorted that, even if our analysis were correct—which is impossible as it conflicts with the received view—it still concerns only half a dozen cases. True, but then even a single counterexample —e.g. the one of the hydrogen atom in intergalactic space—would suffice to ruin the Copenhagen thesis that every quantum-theoretical expression concerns some block object-apparatus-observer. Secondly, the examples above are not pathological rarities but typical quantum-theoretical problems. Thirdly, it is possible to formulate the whole of elementary quantum mechanics, in all generality and from scratch, as a theory concerning (analysable) pairs microsystem-environment.[7], [8] Similar formulations of other members of QT are presumably possible.

And what about the quantum theories of measurement? Do not they show that it is impossible to separate the object from the observation set-up? Let the following suffice here as an answer: detailed arguments are presented elsewhere.[1]

(i) None of the real problems encountered in QT (e.g. the problems mentioned in section 1) require a quantum theory of measurement.

Which is just as well, because there is no such thing as *the* quantum theory of measurement: there are several such theories and none of them is sufficiently developed to tackle actual physical problems.

(ii) A genuine (not phoney) theory of measurement should satisfy at least the following conditions:

(a) it should be *consistent* with the basic QT or with some extension of it;

(b) it should be strictly *physical*: it should not contain any extraphysical (e.g. psychological) concept—which is to say that it should concern itself exclusively with the physical aspects of measurement;

(c) it should be as *specific* as the measurement set-up it purports to account for: a generic theory of 'measurement', i.e. a theory of no measurement in particular, is ghostly, for there is no such thing as the universal (unspecific) measurement device;

(d) it should be *testable*: it should yield predictions other than (but compatible with) those supplied by ordinary QT.

By the first remark, the question of the objectivity of QT can be discussed independently of the quantum theories of measurement. By the second remark, no existing theory of measurement will add anything to the discussion on the interpretation of QT—save confusion. In fact, none of them satisfies all four conditions (a) through (c) above. In particular, von Neumann's account of measurement conflicts with ordinary quantum mechanics, as it assumes that the Schrödinger equation ceases to hold during the measurement process. (Instead, the abrupt projection of the state vector onto one of the eigenstates of the measured observable sets in.) Moreover, that account is shrouded in a psychophysical cloud, so that it fails to meet condition (b). (Actually, since no behavioural variables occur in von Neumann's *formulas*,† that cloud has only an obscuring role: what we have in fact is two physical theories, one of which—the theory of measurement—is never actually used.)

It is possible to discard all the psychological ingredients and build a strictly physical theory of measurement; in fact there are already several such theories. [7], [10], [11], [12], [13], But the question of the use of any such theory remains: indeed, a generic theory of measurement is necessarily unrealistic, for there are no general measurements. Now, a generic or unspecific theory is, strictly speaking, untestable: every

† von Neumann[3] admits himself that the subject 'remains outside the calculation': it is, then, supernumerary.

confrontation of a factual theory with empirical evidence requires, at the very least, the adjunction of subsidiary assumptions—e.g. a specification of the Hamiltonian, or of the density matrix—and the adjunction of a set of data, such as boundary conditions and parameter values.[14] In short, the failure of measurement theories to comply with the condition (c) of specificity renders them unable to meet the condition (d) of testability. Hence the existing quantum theories of measurement are curios rather than well-built and fruitful theories. This is why experimentalists seem to have taken no notice of them.

The main reason such theories of measurement are sometimes taken seriously (by some mathematicians and some philosophers) is that they are supposed to provide both the content and the test of QT. This belief is, alas, mistaken. First, the content of QT is specified jointly by all its basic assumptions, as well as by the experimentalist's reports on the behaviour of QT in the laboratory. Secondly, while QT has been subjected to thousands upon thousands of tests, no quantum theory of measurement has so far been tested—which helps explain why these theories lead a sheltered life in the pages of books and journals. This is an aberration: nowhere else in science are idle theories paid any attention to. And this aberration has a philosophical root that must be extracted because it is a continual source of confusion. This root is operationalism—the tenet that every scientific theory should concern operations of some sort or other. Since the obituary to this philosophy has been written several times,† we shall not pause to fight it here.

Drop operationalism—that belated offspring of naive anthropomorphism and well-meaning but short-sighted empiricism—and the whole issue is clarified. One realizes then that a theory may not refer to the same facts that constitute an evidence for or against it: one sees that reference and evidence must not be confused with one another.‡ One realizes also that, for a physical theory to be testable in a physical laboratory (rather than in a psychological one), it must not concern one's own private life (e.g. the act of reading a meter), but it must be about public objects, even if such objects are accessible only indirectly, i.e. with the help of further (e.g. classical) theories. Even psychologists aim at building thoroughly objective theories,

---

† For criticisms of operationalism see Hempel[15] and Bunge.[16]
‡ The confusion between the semantic concept of reference and the methodological concept of evidence, which characterizes operationalism has been criticized by Feigl.[17] See also Hempel.[15]

i.e. hypothetico-deductive systems that ought to hold no matter who uses them, and that ought to refer to organisms under the most varied circumstances, in particular when the experimenter is not looking at them. If this is true of the science of mind, why should it not hold, *a fortiori*, for the science of matter? Just because of a philosophical prejudice that has long since been given up by philosophers?

4. Concluding remarks. It is reassuring to find that a semantic analysis of QT yields exactly the same results as a meaning analysis of classical physical theories, namely

(i) that the reference class of any quantum theory is a set of physical systems, or of pairs (or triples, or in general n-tuples) of physical systems; and

(ii) that it is unnecessary and moreover misleading to ascribe such systems any mental properties, even worse to fuse them with the subject—particularly since we know so much more about atoms than about observers.

The good old concept of physical object—of a thing that exists whether or not it is being perceived or conceived—is thus vindicated: QT is about physical objects. This is why it counts as a physical theory; and this is why it can tell us nothing about ourselves—e.g. our sensations and expectations.

The explicit indication of the (intended or hypothesized) reference class of any QT—or of any other scientific theory for that matter—can and should be left to one or more postulates of the theory. These must be assumptions proper rather than arbitrary semantical rules.†
The referential interpretation of the symbols must tally with the mathematical structure of the concepts they stand for: otherwise the interpretation would not be strict but adventitious, i.e. groundless. Thus, if a state vector depends only on a particle coordinate, it will be improper (groundless) to claim that it also concerns a nondescript apparatus—or the cat next door for that matter. For a variable to be such it must make some difference to the value of the function concerned. If it does not, it will be a ghost variable: one that meddles but does not work.

What has all this got to do with the attempts to do better than QT? It can be argued that, before an attempt is made to go beyond QT, the present QT should be properly understood. And such an understanding requires, among other things, the elimination from its formulation of all the ghost variables concerning idle measurement

---

† For the hypothetical status of many of the semantical 'rules' see Bunge.[18]

set-ups and idle observers. Likewise, special relativity might not have been born had absolute space not been extracted from classical mechanics, and had the ether not been expelled from electromagnetic theory.

Surely the two tasks—the one of reshaping the old theory and the one of inventing a new one—can be carried out side by side. But not in total independence from one another, as every new theory grows in the womb of some of the existing theories. The better the shape of the parents, the easier the conception will be. And here, just as in every other theoretical endeavour, the very first question to be asked is: *what are we talking about?* i.e. what is, or should be, the reference class of our theory? (And this is where semantic analysis comes in.) For, if we do not know the answer, then we shall not be able to apply and test our theory: it is one thing to test a theory purporting to account for microsystems and another to test a theory concerning our states of mind. If, on the other hand, we claim that a given theory concerns not only microsystems but also ourselves, then we must justify this claim by exhibiting all the variables and the formulas that demonstrably point to ourselves. If we are serious about it we shall have to bring in the whole of psychology: we shall thus establish the science of quantum psychology. But if we are more modest and believe that microphysics should go on calculating and explaining energy levels, scattering cross sections, and the like, then let us formulate our theories in a strictly physical way, without any pseudo-philosophical deadwood. That is, if this is our aim then let us keep the realist epistemology that helped Galilei establish a science free from anthropomorphic philosophies. But, whatever way we choose, let us be clear what we are talking about: otherwise our very sanity will be questioned.

REFERENCES

(1) Bunge, M. What are physical theories about?, *Studies in the Philosophy of Science* (editor N. Rescher), Monograph No. 3 of the *American Philosophical Quarterly* (Oxford: Blackwell, 1969).
(2) Heisenberg, W. *Daedalus, Boston, Mass.* (1958) **87**, 95.
(3) Heisenberg, W. *Physics and Philosophy* (New York: Harper and Brothers, 1958), 128 ff. and *passim*.
(4) Bohr, N. *Erkenntnis* (1936) **6**, 293.
(5) Bohr, N. *Atomic Physics and Human Knowledge* (New York: Wiley, 1958), 72 and *passim*.
(6) Feyerabend, P. K. *Philosophy Sci.* (1968) **35**, 309; (1969) **36**, 82.

(7) Bunge, M. *Foundations of Physics* (New York: Springer-Verlag, 1967), chap. 7.

(8) Bunge, M. A ghost-free axiomatization of quantum mechanics, *Quantum Theory and Reality* (editor M. Bunge; New York: Springer-Verlag, 1967).

(9) von Neumann, J. *Mathematische Grundlagen der Quantenmechanik* (Berlin: Springer-Verlag, 1932), 224 and 234.

(10) Danieri, A., Loinger, A. and Prosperi, G. M. *Nucl. Phys.* (1962) **33**, 297.

(11) Bohm, D. and Bub, J. *Rev. mod. Phys.* (1966) **38**, 453.

(12) Groenewold, H. J. Foundations of quantum theory (Preprint of the Institute for Theoretical Physics, Groningen University, 1968).

(13) Krips, H. Theory of measurement (Preprint of the Department of Mathematical Physics, University of Adelaide, 1969).

(14) Bunge, M. Models in theoretical science. *Proc. XIVth International Congress of Philosophy III* (Wien: Herder, 1969).

(15) Hempel, C. G. *Aspects of Scientific Explanation* (New York: Free Press, 1965).

(16) Bunge, M. *Scientific Research* (2 vols.; New York: Springer-Verlag, 1967).

(17) Feigl, H. *The 'Mental' and the 'Physical'* (Minneapolis: University of Minnesota Press, 1967).

(18) Bunge, M. *Rev. mod. Phys.* (1967) **39**, 463.

THE INCOMPLETENESS OF QUANTUM
MECHANICS OR
THE EMPEROR'S MISSING CLOTHES

H. R. POST

The first question to be decided is whether there are problems of general philosophic import that have arisen for the first time in quantum mechanics.† The answer to this question, in my view, is yes. We need only point, for instance, to the problem of incompatible parameters (see section (3) below), a problem that has only arisen at the fundamental level in quantum mechanics.

A scientific theory necessarily involves interpretation in addition to any formalism. Quantum mechanics is not captured by any treatment that confines itself to a discussion of one of its formal representations. Thus we might spuriously call quantum mechanics deterministic on the grounds that the Schrödinger formalism is a partial differential equation with unique solutions (given the explicitly admitted initial and boundary conditions); however, the Born *interpretation* of such a solution is essential to any discussion of the *theory*.

Quantum mechanics differs in several essential points from classical mechanics.

(1) It is a global theory in that it does not draw a sharp separation between an objective (material) world and the fact of (instrumental) observation.‡ Quantum mechanics embraces the experimental apparatus and outcome; in this it differs from Newton's mechanics whose basic statements are objective, and do not relate to apparatus (the instrumental theories in Newton's scheme being separate and separable). In *this* sense, Newton's theory is less global, and hence

---

† In looking for philosophic problems peculiar to quantum mechanics, I exclude such points as the fact that the Schrödinger wave function is more than three-dimensional and is not, therefore, to be interpreted as anything like an ordinary charge distribution, in the sense in which Schrödinger originally hoped it might be interpreted. Once we adopt the Born probability interpretation, it is no more surprising that a two-particle wave function is six-dimensional than it is that the description of the location of two cows in a field is four-dimensional.

Nor is the fact novel that we introduce in the equation of motion a function ψ which is only interpreted by way of $\psi\psi^*$: the classical electric field cannot be measured linearly either.

‡ For an extension of this idea, see Everett.[1]

275

less complete than quantum mechanics: the fact of measurement, of observation is external to the theory.†

(2) Quantum mechanics does not present us with equations of motion in the classical sense. There is no continuity. As Weizsäcker points out, an analysis of classical physics including thermodynamics already establishes that it cannot be a consistent account in terms of the continuous motion of point particles We need a new epistemology to replace equations of continuous motion of real objective point particles by laws linking one state of information to another state of information.

The new epistemology of quantum mechanics has at least two consequences, in my view.

(a) It introduces subjectivity by way of the Born probability interpretation, in the same way that the parameter temperature is, in this 'relative probability' view, a statement of coarse-grained information, of subjective ignorance incompatible with a statement of detailed velocities. 'Was ich *nicht* weiss, macht mich heiss.'

(b) It reduces the physical description to direct information on the results of measurements. This suggests mathematical representation of 'fundamental physics' (such as Bastin's or Atkin's) in the form of statements of a, preferably small, finite number of alternative possibilities. If quantum mechanics is taken seriously, there are no models in the sense of models suggesting further predicates, *at all*. We have what claims to be a complete description in terms of a finite number of predicates. The decision to take quantum mechanics that seriously would, in my view, be premature, since quantum mechanics *is* incomplete at present in a sense to be discussed below.

The problem we are discussing here involves the reduction of wave-packets by observation (Schrödinger—cat; α-decay). It is the problem frequently discussed under the title of 'theory of measurement' (which in my view is a misnomer, since this theory includes any kind of observation, not only measurement in the strict sense of mapping onto the real numbers).

It seemed to be universally agreed, and I certainly agree, that measurement and indeed recorded observation, essentially involve a measure of irreversibility, even if only of quasi-irreversibility, in the sense of a long recursion period, involving a sufficiently complex system composed of apparatus and/or observer. I do not agree with Weizsäcker that this irreversibility requires a mystic primitive notion

† Newton's theory might perhaps be supplemented by a mechanistic theory of the observer etc., thus restoring to it the universal global character of early atomism.

of time. Time should simply be considered a reducible parameter denoting change. In other words, I do not believe it either necessary or desirable to introduce thermodynamics as a primary theory.

(3) The most profound novelty of philosophic import introduced by quantum mechanics is the notion of incompatible parameters: either parameter may be measured as the outcome of an experiment conducted on a system; either parameter may meaningfully be used in the description of a system; but not both. This makes a realistic interpretation of the quantum mechanical description of a system impossible.

The problem of incompatible parameters and the reduction of wave-packets is reflected in the so-called paradoxes as posed by Einstein–Rosen–Podolsky, Frisch, Dicke and Wittke, and Bohm. These 'paradoxes' are not logical paradoxes, but certainly run counter to our intuition in introducing instantaneous action at a distance, and it may not be surprising that relativity is needed to deal with them. The objection to the quantum mechanical consequences of the thought experiments proposed may be described as an appeal to the 'cluster principle' as used in the discussion of para-particles.[2]

(4) The formalism introduces a *distribution* of values of one parameter associated with a well-defined value of the other parameter. Such a distribution or field is alien to classical mechanics.

(5) It is therefore not, perhaps, surprising that the correspondence programme of reducing classical mechanics to quantum mechanics has not so far succeeded. The task of treating a path, or a δ-function, as a limiting case of a well-behaved ψ-function, or a superposition of eigenfunctions with their associated characteristic symmetries etc., may be insoluble. Certainly the time-development of an initially narrowly located ψ-function (spreading of the wave function) does not, in general, match the development of uncertainty due to initial imprecision in classical mechanics.[†3]

The correspondence procedure has to be joined with ad-hoc assumptions, in order to yield results interpretable in classical terms.

Whilst accepting the importance of such a correspondence principle, I do not accept the view held by complementarity philosophers that we essentially require two models (and two languages), one of which is to be classical. I confess that I am not clear whether the people who hold complementarity views are Kantians to the extent of identifying that necessary classical model with Newtonian mechanics.

† This failure of correspondence is discussed by G. Fay in an article to be published shortly in *Acta Physica Hungarica*.

(6) Quantum mechanics is the first *ultimately* probabilistic theory. Unlike classical thermodynamics, which is considered, at least by some schools, reducible to deterministic classical mechanics (e.g. by way of long-term averages etc.), quantum mechanics claims to be an ultimate description.

Whether this claim is essential to the theory, let alone justifiable on grounds external to that theory, is not our primary concern here. What we *are* concerned with is that quantum mechanics, in common with any probabilistic theory, is incomplete.†

(7) We shall define a theory as complete if any statement describing a possible state of affairs in the language of the theory is derivable as a prediction in the theory, i.e. from the laws in conjunction with some initial (and, in some cases, boundary) conditions admitted by the theory.‡

In a scientific theory we are concerned with deriving predictions from the laws with the aid of appropriate initial (and boundary) conditions. Some universal so-called boundary conditions should be assimilated into laws (e.g. certain restrictions defining the set of admissible wave functions in Schrödinger's wave mechanics). There remain 'initial' and/or boundary conditions that are special to the case in hand and are not (in the given theory) assimilated into the laws (nor, according to Wigner will they be so assimilated in any theory). We shall call these 'initial' conditions for short.

A given theory defines its class of initial conditions. Newton's mechanics involves initial conditions specifying the instantaneous position and velocity of the particles. Given the values specifying these initial conditions, we are able to derive (with the aid of the

† We are not, of course, referring to the fact that the theory of quantum mechanics, in common with any theory that contains the arithmetic of natural numbers, is incomplete in the sense that there are statements in the language of the theory such that neither they, nor their negation, can be logically derived from the axioms. This Gödelian difficulty can be overcome, for instance, by joining to a physical theory all true statements of the mathematics used, i.e. usually at least of arithmetic

‡ Whether a theory is complete or incomplete in the sense defined in this paper is independent of the answer to the question whether a theory offers a complete description of reality in the sense discussed by Einstein, Rosen and Podolsky: classical statistical mechanics at least does not suffer from the problem of incompatible parameters, but *is* incomplete in our sense (e.g. it does not predict fluctuations with quantitative certainty). On the other hand, we might imagine a quantum mechanics that does not refer to individual events at all (even by way of interpretation), but only to the outcome of certain experiments which automatically average over a large number of 'individual' events (as measurement of pressure does in classical mechanics), thus avoiding at least that source of incompleteness, whilst still carrying a superstructure of incompatible theoretical terms.

laws, including a force law such as that of gravitation) the predictions of the theory viz. the positions of the particles at any (future) time.

Quantum mechanics, being a scientific theory, has predictive power. However, these predictions are only probabilistic. We cannot predict with certainty the exact position in which a particle will be found after a finite time even if we are given the optimum, maximum relevant initial information the theory allows.

It is nothing but a verbal evasion to point out that the ψ-function in Schrödinger's equation is determined precisely at any future time, given its initial form. The ψ-function only allows us to derive probabilistic conclusions. We do not wish to sweep this flaw under the carpet of (Born) interpretation. The avowed purpose of quantum mechanics is, amongst others, to predict such results as the location at which particles will be detected. It will not do to formulate quantum mechanics as being concerned with probabilistic predictions only. Even probabilistic statements assume a sample space of individual results. A probability statement only acquires meaning insofar as we assign meaning to the statement of individual events.† And quantum mechanical predictions are only refutable (however weakly) if we assign meaning to such statements of individual events.‡

We do not take up any position here concerning the possibility of hidden parameters which might render the theory complete in the predictive sense. It is hardly a controversial statement to say that the introduction of such parameters would require further severe alteration of the present theory. What is important is that the present theory is not a complete predictive scheme. The theory is half-naked; its predictive coverage is incomplete. This fact is not altered by pointing out that there is no successful alternative complete theory at the moment. The theory is incomplete in common with any

† This problem is not advanced in my opinion by changing from the agnosticism here represented to the pagan device of investing the world of phenomena with pervasive wood spirits called propensities.

‡ We do not necessarily subscribe to the view that appeal to classical description is essential for the assignment of meaning.

Nor do we subscribe to the fallacy that it is interference by measurement in itself that necessarily causes uncertainty. It should not be necessary to emphasise that Heisenberg's Uncertainty Principle does not just point out that measurement interferes with what has to be measured but asserts, unlike classical mechanics, that this interference cannot be allowed for (whilst in classical mechanics we can, e.g. allow for the pressure of the calliper gauge).

Heisenberg's *Gedankenexperimente* do not, after all, derive the Uncertainty Principle from classical considerations but, like other *Gedankenexperimente*, articulate the given theory (in this case quantum mechanics).

probabilistic theory, being half-exposed to the slings and arrows of fortune.

Perhaps the clearest example of the predictive incompleteness of quantum mechanics is the disintegration law of radioactivity. To take the example of an atom of a member of a radioactive series: its lifetime is certainly a part of the description in the theory, in fact an essential part of a *test* statement in the theory; but the statement cannot be derived within the theory (this is typical of probability statements; i.e. the mere fact that the theory does not go beyond probability statements renders it incomplete in my sense). We cannot test the frequency of an occurrence without agreeing on individual occurrences.

It is of interest that the basic problem of the independent probability of spontaneous disintegration (leading to the familiar law of unimolecular reaction) was already discussed by Soddy (1904), Schweilder (1906) and Exner (1911). In particular, Schweidler showed that the exponential disintegration law could be derived from the assumption of complete independence of probability of disintegration from age and other parameters. Of course, unless we are Newtonians believing in the derivability of laws from phenomena we do not subscribe to the view that Schweidler *proved* the essential irreducible probabilistic character of radioactivity. (We might, for instance, derive the laws from a model involving collisions between the active atoms and hidden ether molecules.) But he did, undoubtedly, accept the idea of a probabilistic basis and Exner raised this to a general philosophy, statistics being fundamental and determinism only arising by a law of large numbers.

We do not wish here to denounce such a philosophy, we only wish to point out that whether we make a philosophic virtue of it or no, present quantum mechanics is incomplete in the predictive sense discussed above. There is no inconsistency in quantum mechanics, but it would have to be altered severely to be made a complete predictive scheme.

Newton's mechanics is complete in this sense. Any probabilistic theory is not.

To sum up: quantum mechanics certainly admits statements of single events as descriptions of possible states of affairs. Indeed, it depends on them as test statements. They are certainly not meaningless in the theory. The (probabilistic) predictive statements of the theory only gain meaning via the meaning assigned to statements describing individual events. Nevertheless, quantum mechanics is unable to predict single events. Quantum mechanics is, therefore, incomplete.†

The probabilistic interpretation of quantum mechanical predictions is, in fact, inconsistent with predictions of single events, and hence with the requirement of completeness; just as we may not admix a precise description of the motion of all molecules to a thermodynamic description of the corresponding gas.

(8) There is, of course, this difference between the case of classical thermodynamics on the one hand and quantum mechanics on the other. The former is taken to be reducible (*pace* Weizsäcker) at a *deeper* level to a mechanical description (which is inconsistent with the thermodynamical description used at a shallower level; this is a 'complementarity' of modes of explanation which arises generally from the requirement of explaining the known in terms of the unknown, the changing in terms of the unchanging, secondary in terms of primary qualities, and in general any feature in terms not involving that feature).

Whether quantum mechanics will be reduced in this way is an open question. It is clear that the new theory would be inconsistent with present quantum mechanics.

In general, our sympathies should be on the side of those who attempt to reduce one theory to another theory that shows certain advantages, such as greater predictive strength. On the other hand, we should applaud the courage of elementary particle theorists who arbitrarily declare their atomic analysis to be ultimate rather than hedge their bets by appealing to some unspecified sub-mechanism.

Philosophically, we learn from a viable scientific theory not the truth, but the possibility exhibited by the theory as a model. We have known for over a century that a probabilistic theory is possible. It is at least as bigoted to conclude from quantum mechanics that probability is fundamental, as it is to conclude from Newton that determinism is necessarily true.

Whether the theory should be altered so as to achieve completeness and whether it will be so altered, are two further separate questions. It may well be that reference to strictly independent individual localised events will disappear from the theory and be replaced by statements relating to collectives of particles linked cosmologically

† This fact is not refuted by pointing out that this incompleteness is a firm feature of the theory (we never predict single events; the Emperor never is completely dressed); that it is impossible to make it complete (there is a shortage of clothes); or that it is undesirable to complete it (the Emperor looks very pretty half naked); moreover, that he is the only Emperor we have got.

(the Pauli principle may be a first step in this direction). This would be a heavy price to pay. In order to become a complete theory, quantum mechanics would either have to give up (possibly at a deeper level) its probabilistic basis, or it would have to give up its appeal to operational interpretation in terms of classical concepts.†

REFERENCES

(1) Everett, H. *Rev. mod. Phys.* (1957) **29**.

(2) Hartle J. B. and Taylor J. R. *Phys. Rev.* (1969) **178**, 2047.

(3) Born M. *Zeitschrift für Physik* (1958) **153**, 372.

(4) Messiah A. *Quantum Mechanics*, vol. I (1965) ch. VIII, §1 and §5.

(5) Temple G. *Nature* (1935) **135**, 957 and **136**, 179.

† Incidentally, quantum mechanics, is not even formally 'complete' in the truncated sense as defined for example by Messiah.[4] There does not exist in general even a set of commuting operators such that all functions of these operators also commute, and such that the commuting operators between them uniquely define the state *according to quantum mechanical formalism* ('completeness').[5]

Anyway, we do not expect a linear formalism such as quantum mechanics to express the truth about an actual universe abounding in such non-linear phenomena as elementary charges, etc.

HOW DOES A PARTICLE GET FROM
A TO *B*?

TED BASTIN

I think it is important to try to see quantum theory as one among different possible theories. I think we must make an effort to assess its successes and failures from a point of view which does not implicitly assume its essential correctness, while yet trying to give proper weight to the different reasons people have for thinking that the general picture presented by quantum theory is the only possible one.

Quantum theory was developed to take into account a certain class of experimental facts—namely those facts which forced on our attention that there exist discrete attributes of the physical world which cannot be incorporated within an essentially continuous classical theory. It seems reasonable to ask how far quantum theory has succeeded in this task.

Of course, the early forms of the theory never attempted to *explain* discreteness in the sense that they could be said to have incorporated both the discrete and the continuous within one theoretical structure. They simply imposed discreteness as a mathematical constraint on the range of values available as allowable experimental results of the measurement of certain physical quantities. (This description is directly applicable to the energies of atomic structures. It covers measurements involving free particles if we take the familiar probabilistic interpretation of the constant in the uncertainty relation between simultaneous momentum and position measurements.)

As quantum theory developed, however, attitudes towards the problem of explaining the discrete values seem gradually to have changed. At present a majority of physicists probably regard the modern form of the quantum theory as a coherent intellectual structure within which both discrete and continuous quantities appear properly related, and consider that modern quantum theory gives us an understanding of the intrusion of discreteness within continuum physics.

I regard Bohr's complementarity doctrine also as a theory whose first and essential function is to explain the existence of atomicity in circumstances where only concepts of continuity physics are considered operationally well-defined. The finite value of Planck's constant is the essential connecting link between the continuum concepts and the explanation of discreteness.

This account of Bohr's complementarity seems unacceptable to most: there is however, one problem, in the context of the explanation of the discrete, which no physicist would claim to have solved. This is the calculation of the values of the atomic and other basic physical constants. A quite elementary understanding of quantum theory may give us the idea that the uncertainty relation specifies a (statistically defined) lower bound on physical measurement to which the constant specifies a numerical value. It then turns out that to specify any such idea as this exactly one must derive absolute units from the familiar dimensionless constants which can be formed as ratios of the dimensional constants. These dimensionless constants can therefore be regarded as parameters which determine the scale of the microscopic phenomena in terms of the cosmological. This is how we see them if we think of a continuum physics with constants mathematically imposed from outside the theory. If we think in terms of a closed or self-sufficient continuum physics then they become parameters which can be interpreted as coupling constants or interaction constants which specify the relative strengths of the fundamental fields of physics. These constants have for a long time been a source of interest to those who would like to imagine the quantal situation from a starting point which does not presuppose the correctness of current quantum theory. The best-known contribution to the history of thinking about these constants was Eddington's conjecture that they are more fundamental than the dimensional constants ($\hbar$, c, and the like) of which they are usually written as ratios, and that they may well originate in hitherto obscure algebraic relationships like group structures. This contention was opposed by Dirac,[2] who argued that no importance need be attached to the values of the interaction constants since they might change with time, in which case these values at any particular epoch would not be significant.

The difference in outlook which underlies such different evaluations of the significance of the values of the interaction constants can again be reduced to a difference of opinion as to whether quantum theory really explains the existence of discrete magnitudes. If it does explain them then there remains a problem which may or may not be soluble within quantum theory itself—namely, the calculation of the actual numerical value of the constants though there can be different opinions as to the seriousness of this problem's being left unsolved. If, on the other hand, quantum theory has not given an adequate explanation of discreteness itself, then the constants constitute a much more pressing problem, for one has—if one is concerned at all with the generality in physical theory—to construct first an existence theorem:—namely that definite interaction constants exist (i.e. that they have some value rather than no value) and only then separately to calculate the values they actually have. An argument like Dirac's—accordingly—is only cogent at all if one is already independently assured theoretically of the existence of atomicity in the world.

I have now made a case for thinking that quantum theory has not explained the most basic fact it set out to deal with. Adopting this view, accordingly, for the sake of argument, I must first discuss the nature of its success. I shall assume—to put a complex position in a sharp, unambiguous, if crude way—that quantum theory has its main area of undisputed success in the theory of atomic spectra and in solving problems (such as those of the theory of solids) which arise as fairly natural extensions of that theory.† I argue, moreover, that in this area the discoveries made by quantum theory have been discoveries of *combinatorial relations*, and of predictive schemes expressed in terms of such relations. Where there has been dynamics, in the strict sense, it has been imported from classical ways of thinking, and a way of working has been established in which a rather uneasy association of combinatorial scheme with classical type dynamics has become the rule, with a generalized faith in the unitary nature of physical explanation to serve in place of any real synthesis of these two fundamentally different kinds of thinking. Hence we think in classical concepts which presuppose indefinite divisibility of material, and express the presupposition in the mathematical representation of the space and time continua, and yet we work formally with a discrete theory. The difficulties produced by this situation were discussed, in one way or another, in most of the papers in the colloquium. I shall discuss the situation as it concerns the concept of *particle path* or *trajectory*, because this fundamental classical notion—though simple—presents the characteristic difficulties that will be encountered by any combinatorial theory devoid of dynamics.

In a standard text—Chapter 8 of their *Quantum Mechanics— Non-Relativistic Theory*—Landau and Lifschitz [1] use the mechanics of a particle moving in a straight line to introduce and demonstrate the essential difference between quantum theory and classical dynamics. In the former, the path of the particle becomes progressively smoother as more observations are made of the particle. In the

† C. W. Kilmister's paper 'Beyond what?' in this book, takes a similar line.

quantum-theoretical case things are different; the observations generate a swarm of points rather than a progressively more determinate line which could be interpreted as 'the path of the particle' (fig. 1). In this text, this difference is made the experimental foundation upon which a mathematical superstructure is built. Quantum

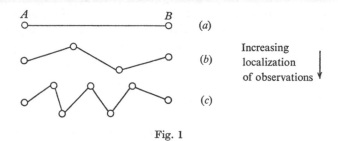

Fig. 1

theory has to kick off somewhere and this is quite a good place; it is not as neat as Dirac's [2] basing the mathematical superposition principle on the behaviour of a composite photon path in a Jamin interferometer, but it is a considerably more useful approach when we are up against problems raised by high energy particles.

When we think about the concept of the path of a quantum particle in the way suggested by Landau and Lifschitz, it is very tempting to idealize the phenomenon by insisting that except where we already have an observation of the position of the particle we can have no knowledge of any sort of the whereabouts of that particle. A given particle process—in this idealized way of looking at things—is that process and no other. Thus, for example, to obtain a 'repetition' of that process with one additional piece of information added (say another collision or disintegration) may require a quite prohibitive increase in the difficulty of the experimental arrangements that are necessary to give a reasonable chance of securing the phenomenon, as well as a quite different mathematical procedure for handling the resulting experimental information. This idealization is one that is becoming more familiar as the operationally realistic approach in the case of high energy particles.

This aspect of the quantum picture is also stressed by Feynman[3] in his presentation of fundamentals, even for low energies. Discussing the two-slit experiment, Feynman observes: 'Now we are not allowed to ask which slit the electron went through unless we actually set up a device to determine whether or not it did. *But then we would be considering a different process.*' (Feynman's italics).

I shall call the idealization that interpolation or extrapolation of points can only have significance on the basis of new experiments being conducted to define each new point, *the quantum idealization*. It is tempting to maintain that the quantum idealization is the whole story. However no physicist ever behaves as though it really is. The physicist always in fact relies on there being a ghostly form of the old classical notion of continuity of path somewhere in the background. For example, suppose we have three cloud chambers A, B, C,

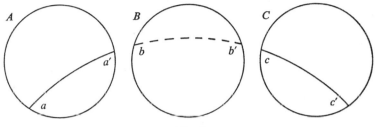

Fig. 2

arranged as shown in fig. 2, and suppose that photographs are taken simultaneously of A and C but not of B. And suppose that the 'path of a particle' is observed at aa' and at cc'. The quantum idealization would require us to say that we had no justification whatever for asserting that if we had photographed B at the same time as we photographed A and C, then we should have got a track bb' (shown dotted). Obviously what the quantum idealization dictates in this sort of case is unreasonable, and physicists are right in their general practice, but to understand how there can be a *limited degree of applicability* of the classical picture of the continuous path is very difficult.

It is possible to take a rigid 'ensemble' view of particle path, according to which the 'particle' is completely undefined for single (or even few) observations. A logically consistent picture can be achieved this way but then all interest shifts to the question why ensembles of single observations so cohere as to provide the appearance of a particle trajectory. I should regard the ensemble *façon de parler* as pointless in the absence of a detailed theory which accounted for this coherence.

Moreover, that Landau and Lifschitz present a *sequence* of cases, makes no difference to my contention that there is more justification for the classical picture than the quantum idealization can provide. The particle path may become progressively fuzzier as greater localization is demanded, but even with high localization a sort of Milky Way of points is left, which has to be explained *somehow*. If one takes the quantum-theoretical position literally then the surprising thing is not that 'particle path' becomes an indefinite concept but that there is any sense at all to be attached to it. Presumably quantum theorists believe that if the solution of the relevant equations could be pursued in sufficient detail then the trajectory-like distribution of points would emerge. This belief however is a pure act of faith and in fact is extremely implausible unless the trajectory is inserted into the theory at some point. In fact there isn't a specifically quantum-theoretical concept of the particle path to be found anywhere.

I now consider the question: what is the *minimum* that quantum theory has to take over from classical thinking to get a particle trajectory? To this end I shall look for a formalized and non-intuitive presentation of the classical concept of particle path; then and only then, can we decide whether what we take over is consistent with the principles of quantum theory or not. Our first essay in this direction might naturally be to consider the definitions of the straight line that are provided by classical mathematical analysis. Does, for example, the Dedekind cut satisfy our requirements? It isn't much good. The operations that are imagined in defining continuity in the Dedekind manner don't seem to have anything to do with the considerations that the classical physicist invokes when he wants to say what he means and what he does not mean by a particle having moved from *A* to *B*.

An account of continuity that is operationally more adequate to the physicist's underlying problem is that used by L. E. J. Brouwer, for whom the continuity of a line required in the first place the possibility of indefinite interpolation of points. (A view which was part of the general constructivist philosophy of Brouwer in which one could significantly speak of a mathematical entity if and only if one had specified an algorithm for constructing that entity.) In this case new points must be generated by some finite process, and what is needed for continuity is to know that there is no limit to the process of selecting two points and interpolating a new one between them (i.e. adding a new point in a given order to the existing set).

We can now list the main requirements necessary to define a sufficient degree of classical continuity in a quantum-theoretical set in order to provide a realistic particle path.

1. The points must be *discriminable*. One must be able to know one from another. (In the classical idea of a line, of course, one would always be able to make measurements independently of the definition of the points which would settle at once the discriminability of the points.)

2. The points will have to be *ordered*. That is to say each point will have to have a definite successor and a definite predecessor at any given stage of building up of the line in order that meaning can be attached to the instruction 'interpolate a point between two existing ones'.

3. There must be a 'topological cohesion' of the points. It is this property of topological cohesion which Landau and Lifschitz point out degenerates as more and more precision is demanded in the localization of points.

Requirement number 3 is intuitive and it is not clear how to express it exactly in general. However, there exists one special case in which it can be given a clear meaning. This is the case of a 'space of potential infinity' in which a construction rule is defined for new points and in which experimental discovery or 'observation' of the new points at this abstract level is identical with the construction process. In this special case the smoothness of the curve or 'tendency of the points to keep together' is not something over and above the discriminating and ordering of the points, but is the same process. It is not to be deduced from these remarks that all trajectories will be smooth—only that the concept of cohesion has been defined and further discussion of detailed cases (like the sequence I have quoted from Landau and Lifschitz) will be possible to describe different sorts of cohesion.

My paper in Part v is an application of a theory which uses a 'space of potential infinity' of precisely the sort that I have just described and the foregoing note really provides the general arguments to justify such a radical attempt.

REFERENCES

(1) Landau, L. D. and Lifschitz, E. M. *Quantum Mechanics—Non Relativistic Theory* (London: Pergamon, 1958).
(2) Dirac, P. A. M. *Quantum Mechanics* (Oxford, any edition).
(3) Feynman, R. P. *Theory of Fundamental Processes* (Benjamin, 1961).

INFORMATIONAL GENERALIZATION
OF ENTROPY IN PHYSICS

JEROME ROTHSTEIN

1. Introduction. The theses here asserted are:

(1) An informational generalization of entropy exists which brings measurement (more generally interaction) within the framework of thermodynamics; irreversibility of measurement becomes an aspect of the second law.

(2) The operational viewpoint is equivalent to adopting a consistent informational language; organization is a concept mathematically equivalent to information, and theory organizes actual or potential results of measurement.

(3) In that language the paradoxes of statistical mechanics (Loschmidt, Zermelo, irreversibility, and assorted demons) are seen to be verbal in the sense that vagueness or ambiguity of description hides a change in the specification of the system under discussion which engenders the so-called paradox; some of the paradoxes of quantum mechanics involve similar ideas (e.g. EPR and other aspects of quantum measurement).

(4) The generalized entropy concept and its associated complex of ideas make it clear in principle (a) why the existence of biology is expected rather than of vanishingly small a priori probability, (b) in what sense it is always permissible to introduce hidden variables, and (c) the kinds of circumstance in which it may be more than empty formalism to do so.

(5) Important clues to relativity and quantum mechanics came from thermodynamic and operational analyses; clues to a future relativistic quantum mechanics may come from an informationally extended thermodynamics.

Space limitations forbid detailed justification of these assertions. We therefore omit elements of such justifications which exist in the literature. References for these cases will be indicated by a starred superior figure in parentheses, e.g. [1*]; they should be taken as part of the argument. Unstarred references play conventional roles. As the presentation is still severely compressed, an apology is tendered in advance for the care required of the reader.

2. Information, measurement and irreversibility.

Information in physics is what is obtained from measurement.[1*] It is not a state of mind in physical discussions. In modern communication theory information is conveyed when choices are made from a set of alternatives.[2] Initial uncertainty about those alternatives is described by a Boltzmann-type entropy defined in terms of the *a priori* probabilities of those alternatives. The information conveyed by a communication or interaction of some sort is measured by the entropy reduction entailed by receipt of the message or occurrence of the interaction. It is measured by the difference between *a priori* and *a posteriori* entropies defined over the set of alternatives. These communication theoretical considerations apply to measurement, for a set of possible indications is provided by any measuring apparatus, and carrying out the measurement is precisely the narrowing of choice from the prior set of possible indications to those actually observed. If a set of alternatives is operationally defined for a phase space, with division of phase space into cells corresponding to some contemplated scheme of measurement, then the corresponding informational entropy is identical to conventional statistical entropy except for a constant (Boltzmann's) depending on choice of units. Physical information and its associated entropy reduction, *localized in the system to which the information refers*, can be expressed via specifications or constraints taken as part of the description of a system, or can be obtained from measurement. Measuring a system and thus *finding* it to be in some state is formally equivalent (i.e. the end result is described the same way theoretically) to *preparing* it to be in that state, *specifying* it to be in that state, or *constraining* it in a manner so that it can be in no other state (the state in question can, of course, be mixed).

Maxwell's demon [3],[4*] is but one dramatic expression of this general informational situation.[5*] Also, the irreversibility of measurement in quantum mechanics can now be seen to have a *thermodynamic* origin. Any successor theory to quantum mechanics would have to have a similar irreversibility imposed on its formulation of the measurement process (by a kind of thermodynamic correspondence principle).

The following points are amplified or made explicit either or both because of their importance, and because they may not be obvious from the foregoing.

(1) Information is generated *only* in a transition from the possible to the actual. As the distinction between future and past is often

described as that between the possible or potential (future), and the actual, factual, or historical (past), it can be seen that irreversibility of measurement and a distinguished direction for time are so closely related that it is hard to distinguish between them. We can learn something new tomorrow but not yesterday because information-bearing choices can be made tomorrow, but the choices of yesterday are already made (note that choosing not to choose is itself a choice).

(2) It is hard to give any sense at all to the term 'past' without presupposing memory or a record of some kind. Information-bearing choices have certainly been made when some physical system has undergone a kind of manufacturing or preparation process, in which either (a) some interaction from without has forced a new selection from the set of states in which the system could exist and still be recognized as the same system (this situation is associated with both classical and quantum measurement), or (b), the system has been transformed into one of a class of systems often distinct from the original one (this situation is characteristic of preparing or manu-facturing a system classically, and frequently encountered in quantum measurement or preparation). Making a record requires the prepara-tion or manufacture of a stable object from an initial system suffi-ciently unstable to be 'triggered' by interaction with the entity whose measurement is being recorded. Furthermore, the one actually manu-factured is a member of a set of objects which could possibly have been manufactured by the same interaction using the same initial unstable object. Selection of the actually manufactured object, from those potentially manufacturable under the given conditions, corre-sponds precisely to the selection involved in making the measurement whose recording consists of the manufacturing process discussed. The manufacturing process is clearly irreversible, being driven by the decrease in free energy suffered by the compound system including the means of manufacture. The time arrows of entropy increase, information acquisition and recording, and both subjective and clock times are all necessarily in the same direction.

(3) Reading a record can be viewed as performing a measurement on the physical object which the *record* is, not the physical object about which the record gives testimony. The irreversibility of record reading or transcription is similar in kind to the irreversibility of the recording operation but is independent of it. Repeated readings or transcriptions of a recording, *as well as the first reading*, do not affect the object to which the information in the record refers. This is important for the Einstein–Podolsky–Rosen paradox,[1*] absolves

the observer who looks through the peep-hole of the murder of Schrödinger's cat, and shows why consciousness can be omitted in discussions of physical measurement; records suffice.[6]

(4) Measurement is an irreversible phenomenon, but the converse is false. Measurement is possible if the system in which the irreversible phenomenon occurs is decomposable into interacting subsystems such that one plays the role of the record plus means for preparing the record, while the other, generally called the object of interest or the like, is what the information in the record refers to. The record need not be stable; it need last only long enough to be 'transcribed' by interaction with still another system. This third system can also be an unstable record which interacts with a fourth system, and the chain can be extended to include an arbitrarily long sequence of such systems, the last of which can be a human observer or a stable record. The general *non-measurement* irreversible situation involves a system, the set of whose *subsystems* lack at least one of the following: (a) a distinguished initial member (system of interest), (b) a distinguished final member (record) and (c), a unique sequence of (intermediate) subsystems whose first member interacts with the distinguished member, then with the second member, after which the second interacts with the third, the third with the fourth, and so on in time sequence, the last interacting with the final member, the directly observable (readable) states of the final member being correlated with the (measurable) states of the initial member. Possession of properties (a), (b), and (c) delimit a special class of non-equilibrium systems, potentially able to function as systems in which measurement or preparation of a (sub)system of interest can occur. Performance of the actual measurement is either part of an irreversible approach to equilibrium (if they are 'triggered' metastable systems), or the maintenance of a non-vanishing rate of free energy decrease (as in continuous recording or monitoring by an 'active' system). The latter case is of particular interest in biology, where the metabolic rate of free energy decrease via oxidation of foodstuff or absorption of solar photons pays the thermodynamic price of biosynthesis, reproduction, etc.

(5) There is a *thermodynamic* upper limit on the accuracy of physical measurement set by the entropy increase available for carrying out the measurement. This follows from (4) because the compound system in which measurement or preparation of a (proper) subsystem occurs must undergo an overall entropy increase. But the entropy of the subsystem calculated for its (macro) state before and after the measurement will show a decrease if its state after the

measurement is more sharply defined than before. The second law sets a bound on this decrease in terms of the overall entropy increase possible in the compound system. In general this is accomplished at the price of degrading energy ΔE, say, giving an upper bound for $\Delta S \sim \Delta E/T$, where T is between the final equilibrium temperature of that part of the compound system in which ΔE degrades and the temperature differing most from the final one during the course of the measurement. Only with infinite energy expenditure or attainment of absolute zero could infinite accuracy be obtained. The latter is prohibited by the third law of thermodynamics, which thus outlaws Laplace's demon.[4*, 7*]

3. Informational resolution of physical paradoxes.

In sharp contrast to carrying out an experiment or performing an operation, which is always historical, singular, unique, and leaves entropy increase traces, theory is unhistorical and general, applying to an infinite class of possible cases. The entropy traces are left by the theorist's metabolism or his records and can be ignored. Were they not we would be dealing with psychology and the history or sociology of science and culture or the like rather than with physical science itself. It is easy to forget that the predictions of theory refer not to any specific actual case but to an ensemble of cases of the form 'If a system is prepared in such and such a state then measurements of these parameters will yield the following results'. Other kinds of theoretical statement are even farther removed from experiment, often being entirely within a formal system. Contact with experience usually arises at the end of a sequence of such statements, linked by derivations within the formal system, by means of translation or correspondence rules between symbols in the formal theory and actual observable quantities. From this point of view theory can be regarded, at least in part, as a metalanguage for the operational language of experiment, whose purpose, among others, is to discuss statements in the operational language, relations between them, and the like.

When language ambiguities crop up in physics, as in the disregard of the distinction between theoretical description of an *ensemble* of possible systems and operational description of the results of measuring or preparing an *individual* system, paradoxes can ensue. Typical are those of space, time, statistical mechanics and quantum mechanics.

(1) Continuous space and time are theoretical abstractions from experience with objects whose separation can be measured and

changed by the observer, and with observed events, remembered or recorded. We have considered elsewhere[8*] how one can start from an operational viewpoint and successively build up subjective time as an informational coordinate, followed by subjective space as an addition to the language needed to discuss external events. It turns out that the topology of subjective time must go all the way to the one-dimensional continuum, rather than being discrete, if the language is to be flexible enough to accommodate all possible events without imposing undesirable artificial restrictions on them. Subjective space needs at least two dimensions in order to permit enough freedom for experimentation; this at least $(1 + 2)$-dimensional space-time has each coordinate capable of being put into one-to-one correspondence with the real numbers. The assignment of numbers admits transformations which cannot affect external events, as they are only changes in how events are named. By considering such 'objectivity transformations' in detail, one concludes that part of the minimal language for describing events and operations is a $(1 + 3)$-dimensional continuum.

This is a kind of prologue to relativity in that each observer comes to this conclusion as a kind of self-consistency condition; only if this is the case does one have the basis for introducing an ensemble of equivalent observers. Note that time, introduced in this way, has inescapable unidirectionality and a kind of statistical ensemble nature. To see the latter point more closely, note that from an operational point of view we always deal with a discrete set of events; but to avoid arbitrariness we have to allow for the possibility of interpolated events (those occurring 'between' clock ticks, for example), leading to a discrete dense set. *The continuum results only when one adjoins limits of all possible sequences of interpolated events.* It is an ensemble of labels for *potential* events. The space-time continuum is thus, in a sense, a minimal framework for events without an arbitrary cut-off. It is thus part of the *language* of physics and may contain *more* than 'reality'. The transition from discrete to continuous has thus introduced something with an aroma of hidden variables. Events are 'manufactured' in the same sense measurement of positions and momenta manufacture the corresponding concepts. The question as to whether future developments might require introduction of a fundamental length or time, implying a kind of quantization of space-time, can then be paraphrased as asking whether passage to the continuum (just as in the introduction of particle individuality) has not introduced into the language the ability to say more than is operationally justified. Should this turn out to be the case, I think it will prove convenient to retain the continuum language, but with the introduction of an analogue of symmetrization to throw away the unjustified information implicitly assumed in the language itself. One could hope to avoid divergencies, perhaps, in this way. A clue in this direction is already afforded by the information–measurement–entropy concept; one cannot localize an event in space-time more accurately than the limit set by the entropy available to spend in the measurement process. The thermodynamics of the future may well absorb space-time.

(2) The paradoxes of statistical mechanics appear to be verbal in the sense that either possession is implicitly claimed of information which is not obtainable or it is implicitly assumed that a system has been prepared in some state when it has not been so prepared. Gibbs' paradox is an example of the first kind. Those of Loschmidt and Zermelo are of the second kind, and can also be viewed as confusing individual and ensemble.

The Gibbs paradox arises when the formula for the entropy of mixing of two gases is applied to two samples of the same gas. It disappears when the false assumption is dropped that an operational procedure exists for deciding whether a particular molecule of gas A came from the right or left of a partition originally separating two samples of A at the same temperature and pressure. In the case of gases A and B, with A originally to the left and B to the right, say, chemical analysis is available to make the choice. Molecules of A and B are distinguishable from each other, but molecules are not individuals in the sense that two molecules of type A possess no name tags or serial numbers enabling us to tell them apart throughout their histories. Our linguistic habits lead us to ascribe individuality to objects, but name tags permanently carrying that individuality, for microscopic objects, are 'hidden'. Symmetrization, which lumps into one state all those linguistically distinguishable but operationally indistinguishable states differing only in name tag assignments to similar particles, is simply the formal way to disregard *linguistic* choices with respect to which uncertainty cannot be resolved and which thus can convey no *physical* information. The name tags can be regarded as values of a discrete hidden parameter. Symmetrization is then simply averaging over all values of the parameter, taken with equal probabilities for all values. This differs from the conventional averaging in statistical mechanics only in that the parameter is hidden microscopically, wheareas phase space averaging is done over parameters hidden from a macroscopic, but not microscopic, viewpoint. Quantization implies that measuring an observable hides its conjugate. From an informational point of view the introduction of hidden parameters which are averaged out again is always admissible in an infinitude of ways, but amounts to no more than a change in convention regarding the additive constant in the definition of entropy. Such exercises are not futile if their theory leads to new experiments whose outcomes differ depending on the values of the hidden parameters, i.e. the parameters cease to be hidden. While the theory is being elaborated, but before new experiments are done, it is naturally not clear whether utility or futility will be the outcome, and both have occurred.

The reflection paradox of Loschmidt, originally proposed to refute dynamical explanations of irreversibility, points out that reflecting all trajectories of the particles of a gas, say, for an arbitrary solution of the equation of motion, yields another solution and therefore decreases any function of the dynamical variables which was increasing along the unreflected solution. If entropy has a dynamical explanation this is a paradox, for then entropy is as likely to decrease as increase. Actually the microscopic entropy, by Liouville's theorem, is rigorously constant because phase volume is invariant under the group of contact transformations induced by the dynamical equations. This is informationally sound as the dynamical equations are exact, taking sharply defined states into sharply defined states. The one-to-one transformations they induce are information-preserving. Irreversibility comes not from the dynamics of the microscopic situation but from the operational situation involved in preparing a macroscopic state.[9*] An ensemble of microscopic situations compatible with the macroscopic specification is described by the usual statistical entropy, the representative points filling out a phase volume $\Delta\Omega$ of the total volume Ω, say. While $k \ln (\Omega/\Delta\Omega)$, the initial entropy, is unchanged in the sense that ergodic streaming induced by the equations of motion keeps the representative points always in a volume $\Delta\Omega$, this volume becomes more and more 'filamentary'. According to the alternatives presented by an *arbitrary fixed measuring scheme* the filament threads through more and more cells of the decomposition of phase space representing that scheme. *The entropy describing the informational situation presented by any specified measurement scheme thereby increases.* Physical entropy is connected with the macroscopic informational situation, and increases irreversibly. In the microscopic informational situation the corresponding entropy is always constant. Paradox arises only because the two different kinds of information are confused. We have shown [9*] that where reflecting the dynamical trajectories is operationally possible, as in spin-echo experiments, information loss or entropy increase is suspended, and signal emerging from apparent noise seems like an entropy decrease. In general, of course, preparation of a reflected state from an arbitrary state is not possible operationally. Whenever it is the resulting decrease in entropy of the reflected system is paid for by entropy increase in the preparation means.

Zermelo's paradox (Poincaré recurrence cycle) is handled in a similar way. [9*] The connection between information and entropy suggests the existence of physically undecidable questions whose undecidability is a novel form of the second law. This is the case, examples being the existence of non-self-predictability for a 'well-informed heat engine' (this is close to the problem of free will), the impossibility of determining the origin or ultimate fate of the universe in terms of its dynamical laws and observable state, and of finding causes or purposes in nature (with reasonable meanings of the terms). [7*]

(3) Quantum theory, like classical theory, gives information about states of physical systems, under specified conditions, given information about how the system has been prepared. Predictions or retrodictions are special cases of this. This informational or organizational function of the theory as a whole can be seen through the device of forming the product space $S \times C$ of the manifolds of initially specified and calculated states S and C respectively. As theory maps the former on the latter, i.e. not all couples (S, C) from $S \times C$ are compatible with the theory, the phase volume of the product space, and thus the entropy calculated for it, is less for the subspace compatible with the theory than for the whole product space. This reduction in entropy, for a specified phase space, measures the amount of information, or organization of experience, provided by theory. [5*] The wave function is the mathematical carrier of all the information available about the system, given an operational description of how it was prepared.

Paradoxes of quantum mechanics like that of Einstein, Podolsky, and Rosen, [10] vanish if one adopts this informational interpretation of the wave function. [1*] As soon as one obtains new information about a system, one's *description* of its state, namely the ensemble compatible with the information at hand, changes immediately. The discontinuous change is not instantaneously imposed on the actual system of interest, and there is no need to associate with it a 'real' process in which mysterious waves travel faster than light. The trouble is caused by the fact that the maximal description afforded by quantum mechanics is considered incomplete on a classical picture. It is important to emphasize that this kind of incompleteness is 'outside' of the language structure of quantum mechanics; the most complete description possible in quantum mechanics is represented by a pure quantum state. Completeness is thus demoted to a relative concept in the sense that quantum mechanics is incomplete only with respect to things with which it is incapable of dealing. This kind of incompleteness, however, is characteristic of any theory or language; no theory has ever been able to encompass the whole universe and everything in it, and metalanguages are needed to discuss questions undecidable or meaningless in given languages. One can, therefore, resolve the completeness issue by saying that both Einstein and Bohr are right, the contradiction being only apparent because they are talking of different things.

Einstein's main point really has to do with physical reality, the kind of incompleteness that bothered him corresponding to the fact that an observer could freely choose what parameters to make 'real' without 'affecting' the system. It is this aspect of the paradox that caused Einstein to view current quantum philosophy as too tolerant of solipsism. He would not admit into his concept of reality the possibility that a distant observer, too far from a system to interact with it directly in the times involved in measurement, could make either of two of its physical observables 'real', the other then being instantaneously made 'unreal'. Again, the informational interpretation finds no great difficulty. Choice of a variable to measure merely permits one to select what *kind* of information is to be collected; description of the subsequent state of the system reflects the acquisition of knowledge of that variable. If one chooses to measure a different variable, one naturally gets a different informational situation and thus a different description. Einstein then seizes on this to say that quantum mechanics is a kind of statistical mechanics in which a 'complete' description of a physical system is not possible.

There is justification to this criticism, but as in the witticism of the early days of psychoanalysis, the situation is 'hopeless, but not serious'. *Any* finite amount of information, where there is a continuum of possibilities, leaves room for an ensemble of possible descriptions (i.e. states of real systems) compatible with the information at hand. As no measurement can squeeze this ensemble down to where the representative point is localized in a vanishingly small volume in phase space, we find that classically the same situation obtains as in quantum mechanics. The infinite amount of information needed to achieve phase space localization of the representative point would require infinite entropy increase and thus infinite energy expenditure.

The informational points of greatest interest pointed up by EPR and wave-packet reduction are first, that freedom to do experiments can be admitted only at the price of making theory treat ensembles. The 'ambiguity' in EPR to which Bohr objected corresponds precisely to delaying the complete operational specification of the situation to which the theory is to be applied. Theory can make no predictions until the specification is complete, so Bohr correctly defends quantum theory against any charge of formal incompleteness not applicable to all theories. But EPR's point that the individual system has no real place in theory stands, and applies to all theories with operational foundations.

4. Generalized entropy and biology. Wigner[11] has shown that a purely quantum mechanical description of a self-reproducing system is not to be expected on mathematical grounds. As self-reproduction is essentially the preparation of a (daughter) system within a larger system (parent plus environment), it is clear that irreversibility is inherently involved. But irreversibility is thermodynamical, not dynamical, so that this does not appear to be a new limitation of quantum (or classical) theory.

When a measurement is made two systems are coupled and entropy is spent in a manner serving to increase the information available about the state of one of them. A similar thing happens during the normal metabolism of a living organism. An organism behaves selectively, i.e. chooses a sequence of actions out of a totality of possible actions, and this selectivity is paid for by entropy increase in the surround. Local increase in organization is paid for by global decrease in organization (entropy increase) exported from the locality in question. A cell gets food from the surroundings and organizes it into a replica of itself. In order to do this it must store information permitting the appropriate choices to be made; the stored information is expressed by the imposition and relaxation of constraints on the thermodynamic course of events. The rate of entropy increase is controlled and channelled. Metastability is of the essence, and irreversibility of the overall process (many, if not most, of the individual steps involved may be practically reversible however) is inherent.

It is a challenge to the information-organization-entropy concept to explain how selective behaviour, in accordance with stored information, is possible in principle at a macromolecular level.

We believe that this challenge can now be met successfully, so that one is not far from being able to see that biology really is, in principle, a branch of physics. In briefest outline, the argument is as follows. [12*] A chain molecule can be viewed as a 'string of beads'. The beads are divided among a set of distinguishable categories (e.g. the different amino acid residues in a protein chain, or the different nucleotide bases in DNA and RNA); chemical storage of information is clearly possible in a code in which the different kinds of beads are analogous to letters of an alphabet. As the chain flips around, in a solution let us say, it assumes many configurations. These are a very small subset of the way the beads might be placed if they did not get in each other's way (excluded volume effect). Because of exclusion, not all configurations are mutually accessible from each other by flips through configurational changes permitted when exclusion is taken into account. The set of configurations thus splits up into 'families' of configurations, such that configurations in the same family are mutually accessible under flips, but transitions between families are forbidden. Configuration space thus has a non-ergodic structure, and an ensemble of molecules prepared in a given configuration would be described ultimately in terms of the entropy of the initial family rather than the larger entropy of the set of admissible configurations extending over many families. If there were a way to 'turn a valve' and make a family of configurations accessible from another family, then the decrease in free energy consequent to the valve action could be used to drive a process of which one step consists of a reaction which occurs only when a configuration in the second family is assumed. If we now say that a particular bead sequence constitutes a specific catalyst permitting this kind of valve turning, then we have a unit of controlled action of a sequence of such actions. The sequence eventually adds up to highly selective behaviour. The 'program' is stored in the 'bead message'.

We have followed this general line of research in considerable detail for a simple particular model of a chain molecule. We have been much impressed with the rich structure of possibilities even for the simplest model, and by the ease with which the interesting properties of the model generalize to realistic cases. We feel that the resources of even the simplest model will make it possible to embed automata theory within it (this would be an existence theorem for biology *par excellence*).

This discussion has been information-thermodynamical; quantum mechanics is used only implicitly (as it is in crystallography, for example) to guarantee the stability of matter and to give molecules and molecular subgroups fairly definite sizes. Volume exclusion eventually goes back to the Pauli exclusion principle, on which the periodic table and all of chemistry are based. We thus have another case of where quantum mechanics provides the bricks from which thermodynamics builds the edifice. Mechanics without thermo-dynamics is inadequate here because it cannot handle the irreversible processes accompanying the use of stored information to control sequences of molecular events.

It is of considerable interest to see if behaviour which is 'life-like' in some sense is realizable with purely macroscopic devices. Were this so, and if thermodynamics, as distinguished from pure dynamics, were inherently involved, then one would have extremely strong intuitive grounds for seeing the essential basis for life processes, as for measuring processes, at the thermodynamic, rather than the quantum level. Our thinking about this problem began with the realization that information would be a truly significant generaliza-tion of entropy only if new bodies of experience could be organized by the generalized concept which could not be with the weaker one. This led to consideration of the thermodynamic behaviour of a 'well informed heat engine'.[7*] This was defined as

a machine and power source with measuring apparatus to determine, at least partially, the state of its environment and itself, effector apparatus to perform operations on its environment and on itself, and internal pro-gramming, computers, and control circuitry whereby it can carry out particular patterns of behavior in response to the information input through the measuring apparatus and the information and programs stored in its memory.

The richness in behaviour such a machine can display *in principle* is so enormous that one can conclude that *any* operationally defined behaviour is realizable in principle. We only need know how to program it, build suitable measuring apparatus, control systems, computers, etc. A corollary of the translation then possible between logical systems (programs) and operationally specifiable behaviours is that physically undecidable questions emerge as novel equivalents of the second law. The investigations of model chain molecules grew from an attempt to grasp how such machines could be built on a macromolecular level from simple constituents in a solution.

It is clear that such machines involve constraints, controls, non-ergodicity, metastability, information, organization, measurement, entropy, and the switching on and off of interactions involved in measurements or manufacturing operations (manufacture differs from measurement in that the former involves much larger perturbations than the latter, and the distinction is evanescent in quantum mechanics). The centre of interest is also an open, non-equilibrium system. The study of well informed heat engines can thus be viewed as a considerable generalization of thermodynamics based on a generalized entropy concept. Insofar as models of living systems (or any systems, for that matter) can be handled at all by the methods of physical science, they appear to be approachable through this kind of generalized thermodynamics.

5. Conclusion. Informational generalization of entropy provides a language appropriate to the operational viewpoint. In that language many paradoxes simply dissolve, for the ambiguities inherent in previous ways of talking about them are clearly revealed. Included are famous paradoxes of statistical and quantum mechanics. The crucial role of irreversible phenomena in measurement and for the physical foundation of biology is also clearly shown. The historical importance of operational and thermodynamic considerations for both relativity and quantum mechanics suggests that informationally generalized thermodynamics may play a key role in constructing a satisfactory relativistic quantum theory.

This research was partially supported by the National Science Foundation, Grant No. GN–534,1.

REFERENCES

(1) Rothstein, J. *Science* (1951) **114**, 171.
(2) Shannon, C. E., *Bell System Tech. Journal* (1948) **27**, 379, 623. Or any modern text on information theory.
(3) Szilard, L. *Z. Phys.* (1929) **53**, 840.
(4) Rothstein, J. *Methodos*, no. 42, (1959) **11**, 94.
(5) Rothstein, J. (1957) *Revue int. Philos.* **40**, 211.
(6) For a contrary view see E. Wigner in *The Scientist Speculates* (editor I. J. Good; New York: Capricorn Books, 1965), 284–302.
(7) Rothstein, J. *Philosophy Sci.* (1964) **31**, 40.
(8) Rothstein, J. *Physics Today* (September 1962) 28.
(9) Rothstein, J. *Am. J. Phys.* (1957) **25**, 510.

(10) Einstein, A., Podolsky, B. and Rosen, N. *Phys. Rev.* (1935) **47**, 777.
Bohr, N. *Phys. Rev.* (1935) **48**, 696.
Schrödinger, E. *Naturwissenschaften* (1935) **23**, 787, 823, 844. *Cambr. Phil. Soc.* (1935) **31**, 555. *Cambr. Phil. Soc.* (1936) **32**, 446.

(11) Wigner, E. In *The Logic of Personal Knowledge* (London: Routledge and Kegan Paul, 1961), 231–8.

(12) Rothstein, J. In *Cybernetic Problems in Bionics* (editors H. L. Oestreicher and D. R. Moore; New York: Gordon and Breach, 1968), 229–45.
Rothstein, J. and James, P. *J. appl. Phys.* (1967) **38**, 170.

CAN LIFE EXPLAIN QUANTUM
MECHANICS?

H. H. PATTEE

I would like to try, if not an entirely new path, at least a new detour in approaching the measurement problem in quantum mechanics, as well as the more general problem of how physical description is dependent on the epistemological interpretation of the *matter-symbol* relationship.

Quantum theory has proven to be exceptionally stable, as theories go, and resistant to 'going beyond' which is what the title of this symposium suggests we try to do. I would like to emphasize that my basic interest has not been the foundations of quantum mechanics, but the origin of life. I am looking for a clear physical reason why living matter is so manifestly different from lifeless matter in spite of the evidence that both living and lifeless matter obey the same set of physical laws. I would have been very happy to accept quantum theory as it exists, but the origin and nature of life is unavoidably dependent on the writing and reading of hereditary records at the single molecule level, and on the sharp distinction between the genetic description and the phenotypic construction processes. The physical meaning of a *recording process* in single molecules cannot be analysed without encountering the measurement problem in quantum mechanics, nor can the symbolic aspects of the genetic description be understood without an interpretation of the *matter-symbol* relation at an elementary physical level. In short, the physical distinction between living and non-living matter turns out, as one might reasonably have expected, to depend on the most fundamental physical and epistemological concepts.

The measurement problem as it has recently been attacked using ergodic principles to justify a quantum treatment of classical bodies, may indeed prove to be a helpful path (e.g., reference (5)). On the other hand, there still remains the problem of our apparently unavoidable primary dependence on classical concepts, and what Bohr, calls the 'ordinary language' in which we communicate about the process of measurement. We may perhaps make a simple classification of arguments over the foundations of quantum theory by dividing them according to the relative primacy attributed to the microscopic quantum description as opposed to the 'ordinary language' classical description. Toward one extreme, we have theories which assume the primacy of quantum mechanics and proceed to derive classical behaviour from the ergodic properties of quantum statistics. Toward the other extreme, we have the assumption that the conceptual world cannot be divorced from the way it actually appears to human observers and their measuring devices, and that quantum 'theory' is only a useful algorithm to be followed in certain microscopic situations, rather than a picture of the underlying reality.

I am well aware that my own contribution is not likely to 'solve' such a fundamental problem, especially after great efforts by the most competent physicists have failed to produce agreement. But my primary reason for attacking this problem, as I have said, is simply that I have not been able to avoid it in the context of life. It is because the problem of measurement reappeared entirely *independently* of ordinary physics that I believe there is a chance that this detour may add some insights to the basic difficulties.

Living matter behaves differently from non-living matter. I will put the problem of the origin of life as simply as possible for this discussion. Living matter is distinguished from non-living matter only by its *collective behaviour in the course of time*. We know from the detailed experiments of molecular biology that there is almost certainly no microscopic or short-time reaction or interaction within living cells which does not follow the laws of ordinary physics and chemistry. Many molecular biologists conclude from these experiments that life differs from non-life only because life is very complicated dynamically compared to ordinary physical systems. This may have some quantitative degree of truth, but what is qualitatively exceptional about living matter is not the complexity of the detailed dynamics but the time evolution of *constraints* which harness these motions to execute *simple* collective functions. We recognize this simplicity of function which integrates itself out of extremely complex detailed dynamics as the evolution of hierarchical control.[10] In other words, beginning with a common set of dynamical laws for the microscopic motions, we observe living matter evolving hierarchies of collective order, and non-living matter evolving a collective disorder. Even the 'true believer' in total reductionism must agree that this aspect of living matter is different from non-living matter. Unless this crucial difference is explained in terms of physical laws, no one can claim to have reduced life to physics. Therefore the essential question of the origin of life is to find a physical basis for this clear, empirical dichotomy in the behaviour of matter. In particular we may ask: *what is the simplest set of physical conditions that would allow matter to branch into two pathways—the living and lifeless—but under a single set of microscopic, dynamical laws*?

Events and records of events. It is clear that under infinitely precise initial conditions and strictly deterministic and complete laws of motion the concept of more or less order is meaningless. We know that disorder appears in physical systems only when we assume dispersion in the values of initial conditions or where the variables themselves are defined as statistical averages. These two modes of description of matter—the strictly deterministic and the statistical—provide already one type of 'branching' in physical systems. However, this branching is to a large degree a subjective matter depending on the amount of detail the observer chooses to take into account. In other words, this type of branching has its source in the modes of description used only by a highly evolved observer, and it is difficult to imagine this branching problem arising in the primæval molecules which originated life. We shall see, however, that the origin of records from a deterministic system must also involve a second mode of description. The problem is to first explain how statistical modes arise spontaneously, and second—the difficult part—to explain how the 'vital' statistical mode leads to increasing organization whereas the ordinary statistical description leads to increasing disorganization.

The epistemological position, which I shall assume in this discussion, is that the concept of probability is inseparable from the concept of *measurement* itself. In other words, whereas the idea of a strictly deterministic trajectory is an acceptable abstraction, the concept of an infinitely precise measurement is not. How do we justify this? Simply by the assertion that a measurement must be a *record* of an event and not the event itself. Consequently while a record of an event may be in error, it is unthinkable that the event could itself be in error. The evolution of disorder in collections of inanimate matter may therefore be attributed to the propagation of error in records of initial conditions. The equations of motion remain deterministic and reversible, but any *records* of initial conditions are probabilistic and lose their accuracy or significance irreversibly in the course of time. This concept of error in measurement has been carefully developed by Born [3] and Brillouin. [4]

If you accept loss of records as the source of increasing disorder in the course of time, then it is reasonable that increasing order in the course of time must require the accumulation of records. In biological terminology we describe the recording process as the accumulation of genetic information by natural selection. But this accumulation is now apparent only in highly evolved cells in complex ecosystems. The origin of life problem is to explain how this record accumulation began and why it can survive the universal tendency toward loss of records which occurs in non-living matter. What is the simplest physical system in which a persistent recording process constrains future events? By stating the origin of life problems in this way, it is clear that we need to know more precisely what we mean by the 'simplest recording process'.

What is a record? I believe we must follow the reasonable assumption that the first records were in single molecules, since that is the way they occur in modern cells. The essentially new condition in this origin of life formulation of the recording or measuring problem is that no human observer, no physicist, no philosopher, nor any macroscopic measuring instrument designed by biological organisms can exist in the beginning. We imagine only the motions and inter-actions of the elementary matter, so we can only ask, how does matter record its own behaviour without the intervention of a physicist. Or in other words: How does the motion of matter lead to *records* of these motions?

Someone will probably object that the observer has not really disappeared in this formulation, and that I have only hidden the observer by imagining the existence of an objective recording process which is operationally meaningless, since it is still the human observer who decides when a record has been made. Here I shall simply admit to being a realist, that is, a person who believes that there are aspects of the world which exist independent of this observer's description of the world. I must accept as a meaningful concept supported by empirical evidence that life did not always exist on the earth, and that it was the accumulation and transmission of hereditary records at the molecular level that eventually led, only after billions of years, to observers like myself. But I must therefore add—and this is the central point—that not only matter, but also *records* existed long before physicists started thinking about matter and making large measuring devices.

At first sight this origin of records problem may appear more difficult than the more familiar problem of how physicists now obtain records of elementary motions of matter; but I think this is an illusion created by our familiarity with highly evolved and abstract languages. In effect, all symbols are records in so far as they are marks which stand for something else. We express all our physical laws in terms of symbols without thinking at all about the *physical limits on symbols*. But even though we have evolved from the first recording process, which occurred some billions of years ago, we cannot use the mere passage of time to evade this circle of self-reference of the matter-symbol problem.

In fact, I believe that one source of our difficulties in clarifying fundamental epistemological questions such as the matter-symbol and measurement problems is that we usually start with the most complex evolved systems we know—namely, the symbolic systems created by the brain of man. Our symbolic languages are deceptive because they achieve great *functional* simplicity, like any highly evolved organ of the body, but only through many hierarchical levels of complex integrated dynamical constraints, achieved only after long periods of selective evolution. This is not only true for the externalized written forms of language but applies as well to all the complex internal biological constraints or boundary conditions which had to evolve before written language was possible. It is no wonder we are puzzled, since the symbols and records we talk about are removed from elementary physical systems by billions of years of biological evolution. This is a gap which we should not expect to bridge in one jump.

It is my central idea that the essence of the matter-symbol problem and the measurement or recording problem must appear at the origin of living matter. Symbols and records have existed since life existed. If this view is correct, then it is a more hopeful strategy to begin by asking what we mean by the *first* primitive record rather than question what we mean by our most sophisticated and abstract records. In effect, this strategy forces us to make an objective criterion for a recording process. We must make a distinction between the record of an event and the event itself in terms of the properties of a physical system, without regard to the higher purposes of experimental physicist, philosophers, or logicians. We must be prepared to consider the concept of record without too many sophisticated restrictions. In particular, we must convince ourselves that the time evolution of the physical system with a primitive recording process can lead to a distinctly different path than similar systems in which no such records occur.

The physics of records. What can we mean by a primitive recording process in terms of physical description? In normal usage, the concept of a recording process implies three steps which we may call (1) writing, (2) storage, and (3) reading. The storage of records is usually accomplished by a relatively time-independent set of constraints which forms structures or marks. The smallest record storage we know is embodied in a single molecule like nucleic acid. Largely because of the relative stability and permanence of storage structures, their significance in the total recording process is usually over-emphasized. For example, many investigators think of the origin of life problem primarily in terms of an abiogenic synthesis of the first nucleic acids, as if the appearance of the same type of molecule which now *stores* genetic records were equivalent to originating the crucial writing and reading processes. Actually, as we shall see, it is the physics of *writing* and *reading* records which are the difficult concepts, while the physics of the storage, or of the symbol vehicle, is relatively trivial. In the same way, it is the design and dynamics of the measuring apparatus which is complicated and not the mark or indication of the stored result.

The mathematician Emil Post [13] saw the essence of writing symbols as the preservation in space of time-dependent activity. 'Activity in time' is the source of symbols, but this '. . . activity itself is frozen into spatial properties'. In a simple physical system this writing process could be described as a *selective freezing-out of degrees of freedom*. The key word here is 'selective'. We do not mean freezing-out degrees of freedom as in a phase transition or condensation. These latter processes are statistical events which are not dependent on the detailed motions of the particles involved. Then what can we mean by 'selective' in this context? In what type of physical system does the selection of a new constraint depend on the motions of the system?

Consider the formation of a chemical bond which requires the proper initial positions and velocities of the reactants. Is this what we mean by the 'selective' freezing-out of some degrees of freedom? No, I think not, since initial conditions are by definition the arbitrary or rule-free conditions in our description of a physical system. While it is grammatically correct to say that we have 'selected a number arbitrarily', what this means physically is that we have no rule for selection at all. So we must modify our concept of writing a symbol to exclude selection on the basis of rule-free initial conditions. What we mean by 'writing' is that we must have a *rule of selection to remove degrees of freedom which is independent of initial conditions.* Now in what sense can any dynamical process be independent of initial conditions? This is a question which we must formulate carefully, since the answer to it contains the necessary conditions for a measurement or recording process.

First, we ask the question in terms of the maximum possible detail. We could ask: what properties of the microscopic physical world are invariant with respect to initial conditions? In a strictly detailed sense, the answer we must give is that only the *laws of motion* are invariant to initial conditions. Or as Wigner has put it,[17] the other way around, invariance principles are possible only because we divide the world into two categories: the initial conditions and the laws of nature. Now it is at least logically clear that to the extent that we require by 'writing a symbol' or 'making a measurement' some selective dynamical process which is invariant to initial conditions, we must, in effect, introduce a new 'equation of motion' for the system, and this is clearly contradictory if we have assumed the original equations of motion are complete and deterministic.

All records are statistical. One way out of this contradiction is, as we know, to relinquish the detailed description and, through a postulate of ignorance, define new variables as 'averages' over an extended time interval or over a collection of microscopic degrees of freedom. These statistical dynamical descriptions 'almost do not depend' on the microscopic initial conditions. Using such macroscopic variables, it is at least not contradictory to introduce *nonholonomic* equations of constraint which may be defined as nonintegrable relations between the new macroscopic coordinates and momenta which must be preserved throughout the motion. Such constraints must, of course, have a physical structure to maintain these relations. In effect, then, such artificial 'invariants' of the motion can selectively reduce the number of dynamical degrees of freedom, and therefore can fulfil our condition for *writing*. But in return for our ability to selectively control degrees of freedom in a macroscopic system, we must accept a corresponding *dissipation* so as not to violate the statistical laws of our macroscopic coordinates. That is, for every binary selection or bit recorded, there must be $(\ln 2)\, kT$ of energy dissipated. If this were not the case then we know that we could design non-holonomic 'demons' which would violate the Second Law of Thermodynamics (e.g. references (14) and (8)). Thus the classical concept of writing or recording demands a classical non-holonomic constraint which is inherently statistical in its structure and dissipative in its operation.

The quantum analogue of writing, of course, runs immediately into difficulty since a non-holonomic constraint between the microscopic degrees of freedom could have no physical basis. Even postulating such constraints in quantum mechanics, as we would expect, leads to serious difficulties. Eden has shown how the quantum formalism can be modified to accept non-holonomic conditions, [6] but the constraints do not in general commute with the Hamiltonian, leading to path-dependent values for the wave functions, and 'observables' which have a definite value even though no observation is actually made (i.e. an operator with an effect depending partly on the history of the state on which it operates).

But however we may choose to describe a selection process in physical terms, we must accept the inherent irreversibility of the concept, and hence a relaxation time or some dissipative process must occur in its physical representation. We must therefore conclude that it is logically and physically inconsistent to think of writing a symbol or a record in the strictly deterministic world governed by complete, microscopic laws of motion. The writing of records and symbols is an inherently irreversible classification process, and its physical representation is therefore probabilistic.

The classical ideal of machines and symbols. This interpretation of symbolic systems as inherently probabilistic or incomplete is contrary to our traditional usage. As the classical idea of laws of nature developed from the times of Galileo and Newton, the concept of determinism was almost always associated with the behaviour of machines. The universe, even including living matter, was compared to a gigantic machine, in spite of the fact that the machine is only found as an invention of the most highly evolved living organisms.

The growth of statistical mechanics did not alter the machine analogy since it was generally assumed that the loss of detail was the result of the practical inability of physicists to measure all the degrees of freedom. It was only with the recognition of the inescapable indeterminacy of conjugate variables that the machine analogy to physics broke down. But the machine concept remains an ingrained part of our thought.

We have compounded this trouble by relating our concepts of symbolic precision and computability so closely to the ideal of the classical machine that we are unable to clarify the basis of either formal or mechanical logics. Thus all of our logical and mathematical symbols are assumed to be strictly deterministic both as records as well as in their syntactical rules for manipulation. But in spite of enormous efforts to clarify the foundations of logic and mathematics we still find the intuitive ideas of a 'formal system' of symbols based on the entirely classical concept of a strictly deterministic 'mechanical procedure' (e.g. references (15) and (7)) even though this is a physical impossibility. While we may, like Laplace, imagine an ideal determinism for the detailed events of the universe, it is precisely the assumed completeness and symmetry in time of this dynamic description in which any inherently irreversible process of classification is unimaginable. Thus *all* recording processes, can only *approximate* the ideal of determinism by minimizing the effects of error produced by the fluctuations of irreversible systems. However small we may make this error, it is especially appropriate to remember Planck's warning to use words like 'certain' or 'sure' with great caution: '*For it is clear to everybody that there must be an unfathomable gulf between a probability, however small, and an absolute impossibility.*' [11]

Where does this gulf begin? The apparent paradox of quantum and classical concepts is that on the one hand, we consider all laws and measurements describing the truths of quantum mechanics as recorded and transmitted by the machines and symbols of the classical world, while on the other hand, we find this classical world is only an approximation to the 'truer' quantum world. One established school attempts to evade this paradox by postulating that the 'classical world' always be taken to the limit of ideal determinism when discussing the symbols and records of the quantum world. For example, Bohr requires,[2] '...the unambiguous use of the concepts of classical physics in the analysis of atomic phenomena'. This requirement is fulfilled, Bohr continues, '...by the use, as measuring instruments, of rigid bodies sufficiently heavy to allow a completely classical account of their relative positions and velocities'. Such a postulate implies that the source of deterministic behaviour in nature is in 'ordinary language' and 'heavy machines', while the origin of probabilistic events is in the interaction of these machines with the quantum world. It is very easy then to slip into the false logical conclusion that the quantum world itself is necessarily probabilistic.

A second school attempts to evade the paradox by avoiding the classical world altogether, or at least treating classical concepts as a kind of useful, but basically unreal link between the quantum level and the consciousness of the observer which remains 'shrouded in mystery'. To a greater or lesser degree, this interpretation has been seriously considered by Heisenberg, von Neumann, Wigner,[16] and other founders of modern quantum theory. Since consciousness is not even defined, this attitude does indeed evade the paradox, but replaces it with the mystery of the origin and nature of consciousness.

Without in any way minimizing the intellectual efforts which have produced these acceptable levels of consistency or at least a safe obscurity in the epistemological interpretations of quantum theory, I detect a sense of extremism in these schools of thought which may be born more of the frustration of years of unresolved arguments than of practical strategies for future developments. In effect it appears that each of these interpretations of the quantum-classical paradox introduces a different location for the 'unfathomable gulf' not so much to clarify the problem, but rather, to more cleanly separate two modes of description which otherwise produce unresolvable confusion.

My own approach, as the title of this paper suggests, is more biological than physical. I have said nothing so far that bridges the gulf. I am not proposing so much how to cross the gulf, as *where to place it*. I would place the gulf where it must be narrow—at the origin, between lifeless and living matter. In other words, between physical events and the *most simple*, natural records of these events. Of course our attitude must also be modified to suit the new location. We must relinquish many traditional concepts which have meaning only in the highly evolved world of human life. These include not only 'consciousness' but also the ideal 'classical machine' and the entirely abstract 'symbol'.

I believe that any attempt to describe the origin of life in physical terms will show that the traditional deterministic classical machine analogy to life is used precisely backwards! As Polanyi has so clearly pointed out,[12] all our macroscopic machines and symbolic languages exist only as the product of highly evolved living matter. Classical machines and symbolic systems are in essence *biological* constraints, not physical constraints. It is a simple, but non-trivial observation that classical machines and languages do not occur in the inanimate world. The fact that our classical machines and symbolic systems can be constructed with high accuracy and reliability is not a tribute to classical determinism but to biological ingenuity, or to put it more modestly, it is the *end* product of evolution by natural selection. This evolution does not *begin* with classical languages and classical machines but with the integrated dynamics of *molecular* languages and *molecular* machines. Single molecules function as the writing, storage, and reading constraints in all present living cells and perhaps even in the brain. Furthermore, it is reasonable to expect that some billions of years ago these recording processes were first accomplished with even smaller molecules. In any case, it is certainly clear that the origin of life's records could not have depended on 'rigid bodies sufficiently heavy to allow a completely classical account of their relative positions and velocities'. Let me emphasize that I do not mean that we, as external observers of life, cannot make some useful classical descriptions of life as is now done extensively in the area of 'molecular biology'. What I am saying is that life itself could not exist if it depended on such classical descriptions or on performing its own internal recording processes in this classical way.

Can we experiment with the gulf? How can we expect to make progress with the matter-symbol problem or the measurement problem in quantum mechanics in a scientific sense without experimental criteria for success? Is there such a thing as an 'epistemological experiment' or are epistemological questions inherently metaphysical? Or to put the matter more pragmatically: can any of these interpretations of quantum and classical concepts be distinguished by a measurable effect? There is no doubt that theoretical discussion will continue on such fundamental issues, and no one can say whether experiments will ever resolve the problems. However, it is clear that these problems have taken their present form because of experimental results, and only on the basis of history, it would be a good guess that further experiments will alter the form of these epistemological questions.

A serious difficulty with the strict separation of quantum and classical concepts by a 'classical limit' or a 'consciousness limit' is that no experimental test appears possible. The Complementarity Principle has in fact been represented as an interpretation of the formalism '...covering automatically any procedure of measurement...'.[1] In a totally informal sense, but with similar results, the insistence on including the conscious observer in the physical system leaves any experimental result open to so many 'soft' interpretations that no disproofs could be possible.

Unlike all other interpretations of the measurement problem that I have understood, the placement of the matter-symbol gulf at the origin of life does suggest connections with possible experiments. For reasons which I have more fully explained elsewhere [9], [10] I would expect a selective, macromolecular catalyst or proto-enzyme to exhibit a simple 'writing' process. Briefly stated, such a catalyst selectively removes degrees of freedom by forming a chemical bond many orders of magnitude more rapidly than it would form without the catalyst. Furthermore, the selection process, in keeping with our requirements, is not strongly dependent on the initial position or velocity of the reactants, but only on their inherent structure. This structure is recognized or classified by the catalyst in the sense that of the many collisions with other potential reactants, only the one with the proper structure will trigger the catalytic dynamics which produces the permanent 'record' in the form of a new chemical bond. The dynamical structure of the catalytic molecule may be described in the classical approximation as a non-holonomic constraint which maintains an 'invariant' rule (perhaps over as many as 10^9 catalytic reactions) for producing records of selected collisions with its environment.

In the context of the quantum measurement problem, many questions immediately arise: (1) To what degree is the quantitative dynamics of specific catalysis derivable from classical laws? Specifically, is a random phase assumption (i.e. the Born–Oppenheimer approximation) sufficient to quantitatively account for the selectivity and catalytic power, or does the correlation of phases play an essential role? (2) To what extent do the dissipative or relaxation processes involved in the collisions with reactants and the reset of the catalyst after each reaction involve heat, radiation, or loss of phase correlations? (3) Does the uncertainty relation limit or account for the accuracy of recognition and speed of catalysis? All of these questions are of course inseparable from the problem of the relation of the dynamical to the statistical modes of description, and we must therefore be careful to distinguish time averages which are 'convenient' for us, functioning as calculators or theorists, and those which are 'essential' for the natural functioning of the selective catalyst.

It is possible that the distinction between the dynamical and statistical descriptions will turn out to be an unfathomable gulf in the human brain even though we are looking at the simplest recording molecule; yet I cannot believe it would not be illuminating to know the dynamics and statistics as far as possible in a *natural* recording situation not designed by the human brain. Averages over short time intervals or small numbers of variables may be displeasing to the mathematician, but quite effective for molecular catalysis. There are certainly all levels of selectivity and catalytic rates—all degrees of error, speed, reliability, and permanence in a recording or measurement process which might be modified in an experiment. With such ideas in mind, it should be possible to simplify models of selective catalysts and to design experimental tests on existing tactic catalysts or enzymes which may at least help us approach a clearer understanding of the primary physical conditions for a natural recording process.

REFERENCES

(1) Bohr, N. *Phys. Rev.* (1935) **48**, 696.
(2) Bohr, N. *Essays 1958–1962 on Atomic Physics and Human Knowledge* (New York, London: Interscience Publishers (Wiley), 1963), 3.
(3) Born, M. *J. Phys. Radium, Paris* (1959) **20**, 43.
(4) Brillouin, L. *Science and Information Theory*, second edition (New York: Academic Press, 1962).
(5) Daneri, A., Loinger, A. and Prosperi, G. M. *Nucl. Phys.* (1962) **33**, 297.
(6) Eden, R. J. *Proc. Roy. Soc. (Lond.)* (1951) **205** A, 583.
(7) Gödel, K. In *The Undecidable* (editor M. Davis; Hewlett, New York: Raven Press, 1964), 72.
(8) Landauer, R. *IBM Jl Res. Dev.* (1961) **5**, 183.
(9) Pattee, H. H. In *Towards a Theoretical Biology*, vol. I (editor C. H. Waddington; Edinburgh University Press, 1968), 67.
(10) Pattee, H. H. In *Towards a Theoretical Biology*, vol. III (1970), 117.
(11) Planck, M. *A Survey of Physical Theory* (translated by R. Jones and D. H. Williams; New York: Dover Publications, 1960), 64.
(12) Polanyi, M. *Science* (1968) **160**, 1308.
(13) Post, E. In *The Undecidable* (editor M. Davis; Hewlett, New York: Rowen Press, 1965), 420.
(14) Szilard, L. *Z. Phys.* (1929) **53**, 840.
(15) Turing, A. *Proc. Lond. math. Soc.* (1936) **42**, 230.
(16) Wigner, E. P. *The Monist 48*, No. 2 (1964). Reprinted in *Symmetries and Reflections* (Bloomington and London: Indiana University Press, 1967), 185.
(17) Wigner, E. P. The Nobel Prize Lectures (Elsevier Publishing Co., 1964). Reprinted in *Symmetries and Reflections* (Bloomington and London: Indiana University Press, 1967).

PHENOMENA AND SENSE DATA IN
QUANTUM THEORY: DISCUSSION

D. S. Linney re-opened a discussion which he had with von Weizsäcker which only one or two people overheard. This discussion brought up the epistemological problem of quantum theory in a form which would seem natural to a person trained in the English rather than in the continental school of philosophy.

(LINNEY): There are four or five things it is important to say but first I should like to try and reconstruct what I said to von Weizsäcker. It will be dull for him but it won't be dull for the other people because I think it leads to a different conclusion from what I would have expected. I asked about the meaning of Bohr's statement that we were always, for all time, to keep to a classical description of the instruments in the observer situation. In fact during the colloquium it became clear to me that what was being said was in some ways different from the meaning of Bohr's statement. Certainly what was being used wasn't a description in terms of classical mechanics or classical physics. One reason which I regard as sufficient is that quantum mechanics has taught us that objects should not be treated along with continuity and that you should not insist on using a mathematics of continuity for objects.

(VON WEIZSÄCKER): I see, yes.

(LINNEY): This is really the message of quantum mechanics and I believe it is very important. What is classical mechanics without continuity? Indeed, what is then the point of calling it a classical description?

The other reason which I put to von Weizsäcker came up because he appears to be suggesting that the right language for the description of the results of measurement is what used to be called 'the language of common sense' in England—say by Russell. (In those days the philosophers here used to contrast the language of physics with the language of common sense.) Now, when I think of examples of concepts which appear to be of ordinary experience— things like tables, chairs, etc.—it doesn't seem obvious to me that classical mechanics is the right language to describe them. To put it another way, I think classical mechanics is an extremely abstract and remote language to describe such things as having tea. I should like to see anyone try it—to describe a tea party using classical mechanics.

So I put this dilemma to von Weizsäcker. On the one hand we could take seriously that classical mechanics is the right language because it is as it were a correction of and improvement on the language of ordinary experience for describing things. (In which case, why should I not say that just as classical mechanics' language is an improvement on the language of ordinary people, so quantum mechanics is a further improvement. In this case I would say that the correct language to describe the instruments in the macroscopic situation ought to be quantum mechanics.) Alternatively there is the other horn of the dilemma which I put to von Weizsäcker. It is possible that—as I really believe and I must now declare interest—classical mechanics is not the right language *at all* for describing ordinary experience. It may be a highly abstract theory founded on a particular metaphysics, and that it is only connected very obscurely at certain points with ordinary experience. In this case, why should we not say against Bohr, that the right language for describing this situation—the observer, the instruments, the phenomena—is for example the language of Goethe. This is one example—I could find better ones, but this immediately came to mind.

(BASTIN): You mean the Urblatt language.

(LINNEY): I was thinking of the description of colour.

(VON WEIZSÄCKER): The language of Urphänomen is what you really mean.

(BOHM): Is it possible to characterize that briefly?

(VON WEIZSÄCKER): Later perhaps.

(LINNEY): Was this a correct account of what I said to you?

(VON WEIZSÄCKER): Yes.

(LINNEY): Before I say any more, I think you should hear what von Weizsäcker said in reply to me. Because from what he said there are further difficulties which really lead into what I regard as what still remains an epistemological problem.

(VON WEIZSÄCKER): I don't know if I can repeat myself as exactly as you did, but I'll try in the same spirit. Perhaps I should follow the line you propose in your arguments. From the outset I'd say that—with the possible exception of the bit about Goethe—you are precisely repeating an argument Teller brought forward against Bohr. After Teller had argued for a new quantum theoretical language, Bohr said 'You might as well say that we are not sitting here drinking tea.' Now Teller left out Goethe because that would have been the other horn of the dilemma. Teller stated the one horn in asking the question 'Why should our language not just be the language of quantum theory?'

(LINNEY): Let me ask something. Surely you'd agree that there is a difficulty about continuity?

(VON WEIZSÄCKER): I fully agree. And—leaving Bohr for the moment—I will put forward a statement which I cannot prove but which I have never seen disproved or a counter argument put. A classical mechanics or a classical treatment of continuous motion of continuous bodies is an intrinsic impossibility—if you include thermodynamics.

(LINNEY): I'm quite sure this is true.

(VON WEIZSÄCKER): In a way it is just what history has shown, but I wanted only to say that this is not an historical accident: it is not fortuitous, but is what was to be expected if one thought carefully enough about the problems involved. In this sense I would say that the step quantum theory took most rightly, was that first step of Planck's in the problem of continuous motion in thermodynamical equilibrium. This is precisely the point where the inevitability of quantum theory can be seen, while all other points—even the one from which Bohr started—rely on particular experience. You might say that this particular experience has been interpreted in a particularly unlucky way, and that we ought to do it differently. But this is not so in the thermodynamics of a general field like Maxwell's field, where we have equations which—since they are continuous in the same sense—lead to the very same result. I feel this remark about classical theory to be important because it means that not only has classical physics as a whole proved empirically not to be satisfactory, but that a sufficiently philosophic mind should have been able from the outset to see that classical physics cannot possibly work. And this is very important in my defence of Bohr in order fully to exclude the idea that a classical description is a description by the system of classical physics.

Now you have stated my alternative view: the description is to be given in what is called the language of common sense.

(LINNEY): It might be called that but it is misleading.

(VON WEIZSÄCKER): That I think is why you rightly refer to Goethe. Goethe was a poet and his philosophy has never had the consistency of philosophies like Plato's or Aristotle's or Kant's, nor the consistency which is present in some mathematical systems. But still it is quite clear what he means. Goethe speaks of 'phenomena' and his central concept is that of *Urphänomen* which is a non-reducible phenomenon, and it is precisely his view that to reduce phenomena to thought objects which we consider to be more real than the phenomena, and thus construct the phenomena, is to turn things upside-down. Really we should say, for example, that the phenomenon of white (i.e. original) light looked at through 'das Trübe'...

(FRISCH): You might translate that 'turbid—containing scattering particles'.

(VON WEIZSÄCKER): To Goethe though, it's different. 'Das Trübe' is one of the basic words which are not to be reduced any more—certainly not to hypothetical scattering particles. He says that light—originally white light—looked at through 'das Trübe' becomes blue, and this is a description—in Goethe's view—of a phenomenon which you cannot reduce any more. You can describe it precisely because you can observe it and connect it with other observations, but you cannot reduce it to anything else because if you try to reduce it you just lose the phenomenon and gain nothing. So in this sense Goethe is not far from Mach, except that Goethe connected the *Urphänomen* with a wider range of things. For instance, it is equally an *Urphänomen* that red is exciting—that colour works on what we call the mind. This attribute of colour is equally primitive and must be so because the *Urphänomene* contain what scientists call values as well as facts, and the combination of value and fact is equally an *Urphänomen* which can only be dissolved by artificial methods. Thus when—for example—values disappear and facts remain, science destroys the world, as it is doing now.

I'm speaking my language now, but I think I am following Goethe's line. Otherwise why was Goethe so extremely angry with Newton? He was angry to an extent which caused him to make completely wrong statements and to fail to understand Newton, though in other contexts Goethe was certainly intelligent. His anger is probably connected with Goethe's feeling that this highly refined scientific method of separating certain facts from the rest is used to destroy the unity of life. This is what really lies behind Goethe's attitude.

This is all partly a sidetrack from quantum theory, yet I think if you go far enough in understanding Bohr you will see that Bohr, in quite different language, touches a similar problem.

But now, having said all this, how can I still defend Bohr's view that one must describe experiments in the language of classical physics? I deal with this point fully in my paper, but to put the point in a challenging way let us assume that Teller is completely right—that we must describe nature in quantum-mechanical language. I say that *means* we must describe nature in the language of classical physics. It means that because that is quantum theory. This seems paradoxical, but whether the statement is correct or not, I must try and make it comprehensible in order that we should continue our discussion.

I shall leave out quantum theory for a moment and discuss the connexion between everyday language—common sense language— on the one hand, and the language of classical physics on the other. The step that leads from the one to the other is considered by Bohr to consist (quoting Bohr) in 'the attempt to give an unambiguous description of what we experience'. Then—since commonsense is often much commoner than sense—there is always the difficulty with the commonsense description that we are inclined to admit ambiguity and contradictions: we can't really avoid them. In the way I interpret Bohr, classical description is nothing more and nothing less than the attempt to give as much unambiguity to our everyday description as we possibly can.

(LINNEY): What you are proposing is a Galilean programme.

(VON WEIZSÄCKER): Yes it is Galileo's programme: he was one of the initiators of modern physics. Now consider where this programme gets us: In the seventeenth, eighteenth and nineteenth centuries it got us the completion of what we now call classical physics and in the twentieth century we have gone beyond that in reducing ambiguity and achieving clarity and precision by inventing quantum theory. Why should we stop at the late nineteenth century instead of definitely going into the twentieth? I claim that this is not a question of historical progress only but a question of the structure of phenomena and of reality which has been revealed to us by our constant work in physics through the centuries. Here there is a break. To see this break clearly we must distinguish two meanings for classical physics. First there is the system of classical physics which tries to include everything in an explanation by appealing to a *model* of nature, which considers nature to consist of bodies which strictly obey the laws of classical physics whether they are phenomena or not. This is wrong—can be proved wrong. The other meaning is to use classical language within the realm of phenomena in order to describe them. Here the point is—as Bohr always said—that the quantum of action is precisely the discovery that in order to give an unambiguous description of what we observe in phenomena we are forced, not to renounce classical models, but to renounce models. Nothing like a quantum-mechanical model which replaces classical models and which then would admit of description of nature in terms of an explanation by the quantum-mechanical model, exists. What exists are rules by which we connect observations which we make (in the original sense, looking at phenomena or producing phenomena) and which we describe as precisely as possible—that means in terms of classical physics. The rules connect these observations without use of models and therefore without applying the complete classical system, only leaving us with the application of classical concepts.

(BASTIN): How much do you identify yourself with Bohr on this?

(LINNEY): He was not merely expounding Bohr; he was also asserting the truth of this position.

(VON WEIZSÄCKER): Yes. Only I have now used the term 'classical' in a sense which has turned out to be undefined.

(AHARONOV): How do you get these rules?

(VON WEIZSÄCKER): That's the same question as how you get quantum theory, because quantum theory is the set of these rules. The main method is the traditional empirical method of trial and error, and though it took time we were finally satisfied with what we had got. There may finally be (and this is my personal aim) the possibility of seeing these rules as necessary preconditions of experience, but this is not Bohr, and it is not needed for the discussion of Bohr. For that, it is quite sufficient to speak as we already have with one addition—namely that the term 'classical' is really not well defined. In identifying myself with Bohr in this context, I do so, therefore, fully realizing that I am under an obligation to define clearly what I mean by classical, and I am going to do that.

'Classical' is really a very unluckily chosen expression. I can understand 'classical music' and 'classical art'. In general, by 'classical' we mean that which is very good but is not what *we* are doing, and in this sense classical physics is classical, but this is not a description of its content or its structure—only of our relation to it. The question is, can we replace this word 'classical' by a word or a set of words—a definition—which would describe clearly why I say we are bound to stay within classical concepts. I am going to try to do this, but in so doing I shall be going beyond what Bohr explicitly said.

I start again with Bohr's argument, which is that in classical physics two basic elements of our knowledge go together in a non-complementary way which he calls space-time intuition and causality. I consider this precise—at least far more precise than what used to be said, because I am now entering a field in which I can use arguments which Kant called transcendental. That means I cannot possibly mean by 'experience' something which would not show up within space and time (whatever 'within' may mean here, for prepositions are the worst of things) and on the other hand be connected within itself by rules which can be described by the general principle of causality. This, I think, can be justified as far as our knowledge goes, though I am not wishing to dogmatize or say that our intuition or our concepts may never change. Nevertheless as far as we understand them we would have to say that experiences are in time. *Experience* can be defined as using knowledge of the past for the future; hence experience certainly presupposes time. Again experiences are in space because experience (at least for the physicist) means that he is observing something that he can see or feel or in some way be aware of by his senses. And the most general frame of what we can perceive by our senses we are accustomed to call space.

Also experience must be causally connected, and here I can give as the simplest example, a measuring device. In a measuring device or instrument—say a microscope—I look at one end of the instrument and make conclusions about what happens at the other end. If I couldn't reach such conclusions it wouldn't be a measuring instrument.

(LINNEY): I might be prepared to accept this argument about space and time with the caveat that there is a difficulty about continuity, but why causality? I can see why for Kant things have to be causally connected, but for me and for my generation of philosophical students it is not only not obvious; it is obvious that the opposite of it is true. And I connect this with the existence of quantum mechanics which has taught us, if we were in any doubt before, that it is better—more rational—to start by being indeterministic in general: I should be prepared to accept this also about the measuring instrument. For in spite of what you said, I still think that it is a presupposition of all science that we and our measuring instruments are a part of the world that is being measured, and in that sense no different from it. There may be logical difficulties about saying this which I am prepared to admit but why shouldn't I hold that a proper formulation of quantum mechanics (it is a question of training a generation to get used to these concepts) would ultimately mean that there would be no more classical description.

(VON WEIZSÄCKER): This is just the point. I fully agree with all you said except the last sentence. I believe in quantum theory in the sense that I believe that nature is indeterministic and I also believe that I am part of the world that is described by quantum theory. Bohr always said that too. From this I draw the conclusion about which Bohr was hesitant and about which I think Bohm and Schumacher are quite sceptical, that it should be possible to describe whatever I experience—including myself in principle—according to the laws of quantum theory. (I say in principle because I cannot do that in practice without destroying myself and this certainly has philosophical implications: it is not a meaningless statement.)

However, leaving aside this last point, I fully agree that what happens in the measuring instrument is indeterministic. That means it is not possible to predict everything that happens in the measuring instrument strictly, however precise your initial knowledge may be. But then I assert that this is a situation which is presupposed by Bohr's statement, and does not contradict it, because otherwise there would be no complementarity. The point is that even though I know that the measuring instrument is not precise—is not deterministic—to just that extent it is not a measuring instrument. Moreover, in order to be useful as a measuring instrument I must confine my use of it to those traits of its behaviour which are deterministic, or sufficiently well determined (which is a question of accuracy—as accurate as I care to be or as accurate as is demanded by the degree of predictability). If those conditions were not satisfied we should not regard the instrument as a satisfactory measuring device.

Now we all know that quantum theory is not complete chaos: indeterminacy in quantum theory means predictability of probabilities, and if we have sufficiently large numbers of cases then the probabilities may, in many cases, amount to 1 or 0, and I can use an instrument in which *this* happens as a measuring instrument. If I am concerned about facts which cannot be observed by such a measuring instrument because no measuring instrument is sufficiently deterministic then I simply say that those facts are just unknown to me, and must necessarily be unknown to me.

I did say—in one of the first meetings—that we need to be aware of the fact that physics is *essentially* an approximation. I cannot attach a meaning to the statement that physics is precisely true. And this refers to physics—not to any particular physics. Quantum theory is the first form of physics in which this fact has shown up clearly even to physicists, and this is the fact which Bohr tried to treat by his concept of complementarity—that one must renounce one thing in order to have the other and you cannot have no ambiguity as well as completeness at the same time.

(AHARONOV): I am not sure I understand your point that the measuring device must have something non-ambiguous about it.

(LINNEY): It was the description that had to be non-ambiguous. For the measuring instrument to be a measuring instrument it must be deterministic. If he admits that it is essentially indeterministic—which he didn't—then he has to say that only in the sense that it is *nearly* deterministic is it *nearly* a measuring instrument. In fact—to make a Wittgensteinian remark he would be replacing the word 'measuring instrument' by 'nearly a measuring instrument'.

(VON WEIZSÄCKER): But I don't want to carry out this replacement because that is what we always mean by 'measuring instrument'.

(AHARONOV): But just imagine the following situation: in classical theory we assumed that measuring instruments had no limit to their being perfected, whereas quantum theory teaches us that in other circumstances there *are* limits to perfectability. Now why can't we assume that this assertion will apply to the measuring device itself? : it too will have a limit of perfectability. There will be a *minimum* uncertainty about the connexion between the measuring device and the thing it tries to measure. *No* measurement can be exempt, not only a measurement of two different things.

(VON WEIZSÄCKER): I have no objection to that, only we must try to find out how it fits in with the way I am speaking now.

(AHARANOV): But I am trying to explain how one can develop a new language about the restriction of the measuring device itself. If there is a limit to the perfectability of the connexion of the measuring device with the system, then this is something new about the measuring device that classical theory did not admit, because in the classical theory there was no limit to perfectability.

(VON WEIZSÄCKER): This ambiguity ought to be removed. When I speak of the classical description of the measuring device I explicitly exclude the idea that this merely means the strict application of the classical theory.

(AHARONOV): No, I mean your condition that for something to serve as a measuring device it should be completely deterministic.

(ROTHSTEIN): There's a contradiction in this, because if a measuring device has not complete *in*determinacy in a certain sense which I will define, then it is totally useless as a measuring device. In short, the function of a measuring device is to present us with an indeterminate situation which becomes determinate when a process called a measurement occurs.

(AHARONOV): That is not the issue.

(VON WEIZSÄCKER): That's another sense. But may I use Rothstein's remark to clarify some concepts which I used without clarification. Let's think for a moment about the structure of classical mechanics and classical physics in general. If you say classical physics is deterministic then that again may be an ambiguous term, and I shall try to define it. In classical physics you have laws which are in general formulated as differential equations with respect to time—perhaps even partial differential equations. These laws exhibit, in their form as differential equations, the particular attribute of determining something but not everything. They determine —to put it mathematically—*this* set of solutions of the differential equations as possible Heisenberg states of the system followed through time. But the whole set is possible: that means that every single one of the solutions might be selected according to the given initial conditions, and in this sense the theory is not deterministic because it does not determine the initial conditions. Now of course in the completeness of nature the philosophical determinist may think that the initial conditions are only undetermined owing to our lack of knowledge, and if we knew everything (like Laplace's scheme) then we should have no need of initial conditions: one would just have to fix the conditions once—let us say at minus infinity—and the rest would be determined.

(LINNEY): This requires that we know everything at one instant of time—instantaneously.

(VON WEIZSÄCKER): This, again, is what I call the *system* of classical physics. It is certainly wrong and certainly to be rejected, but the practical use of classical physics can go very far in the sense that we just accept that there are particular situations in which the initial conditions are not known but the applicability of the laws is sufficiently justified. And then we have precisely the situation Rothstein was talking about. If we have a measuring instrument to which we can apply this sort of classical description, there is complete freedom in the initial condition which I do not know. There is complete determinacy in the connexion between the initial condition and the final state which I can observe. For that reason, observing the final state, I can deduce by calculating backwards what the initial condition that I wanted to know, was. That initial condition is defined by the interaction between the measuring device and the object. In this sense determinacy goes together with freedom because of the particular structure of classical laws which do not determine everything.

All this still applies on Bohr's description of quantum-theoretical measurements because of his condition that the description should be unambiguous. 'Unambiguous' here implies that it should be possible to achieve it by deterministic devices. Now Aharonov's point was this. The conditions of unambiguity of the combination of space-time intuition and causality can only be fulfilled within the frame of those forces and descriptions of things which were already available in classical physics. I wasn't familiar with all your examples, but I somehow had a feeling that there might be something like that. Thus I was very glad to know how to describe situations like your rotation by 2π which classically restores the original situation but according to quantum theory may lead to something different which we can observe. Since we can observe it we can describe it in terms which Bohr would have called 'classical'.

Here, however, the meaning of 'classical' needs clarification, and my remark in my first section (which I'm afraid was not really on the level of the present discussion) about thermodynamics can now be strengthened a bit. I had the idea that to deduce classical description (in the sense I am about to try to define) from quantum theory, one may have to impose certain conditions on the quantum-mechanical description which are at least necessary. One of these is irreversibility. (I quite admit—as I did in my paper—that it is not a sufficient condition.) I don't know what other conditions ought to be added to make it quite clear, and I think no one has really solved this task. Still, I think that the papers of Loinger, Prosperi *et al.* are an attempt to go in that direction, and Aharonov has made attempts of this sort. So we may get clearer in the end. Of course all this presupposes that quantum theory is exact and there is quite another possibility of which Bohm has been speaking that quantum theory may turn out to be adapted just to a certain field of our experience, and that we should go beyond quantum theory in the sense of introducing other schemes—other ways of ordering experience. About that I hold views different from Bohm's which we need to discuss separately.

(LINNEY): It seems to me that in what you say there remain (from the point of view of the reflective philosopher) very serious epistemological questions which in my eyes at least are different now from what they were before. Before what you said I had always taken it that the problem was how to combine some non-Machian picture of physics with the comparatively crude facts. Dirac—for example—said 30 or 40 years ago that all we really do is to connect one set of observations with another set of observations by some mathematical formalism which does not itself require any logical interpretation. This is essentially a Machian position.

(VON WEIZSÄCKER): Yes.

(LINNEY): But now if this is not correct and yet if at the same time we wish to assert it (which in a certain sense is Bohr's position) then we have to give some account of what (since we know he was a Kantian) the objects of knowledge are. This is why the other day

I raised what seems to me to be the most critical question for a philosopher: what are these objects that I was accustomed before to hear termed 'state' or 'quantum state'? If this term is strictly speaking inapplicable for the interference reasons then I am in a difficulty about applying the simple categories of 'property' and 'object' which are involved in the concept of quantum number. I am also in a difficulty about how to understand what the observable is. It is also clear—from another discussion I had with von Weizsäcker—that he is not intending to replace the objective description with a subjective description that would talk about our knowledge instead of talking about the universe. Hence the obvious answer that the object of knowledge is a proposition or a judgment or a belief is no good. What are the objects, therefore, whose properties or whose states are being investigated?

This is the metaphysical question, and it is connected with the epistemological question because in the ordinary language it is quite possible to say that in some sense there are objects, or if you like, particles. They may be abstract, but we can talk about them in subject-predicate language: we can use very simple concepts which occur in classical physics and in quantum physics and in common-sense language. Incidentally of course it's the kind of language which leads naturally to very simple classical logic, because there will always only be two values: you simply ask whether something has a given property or not.

It seems to me that the only constructive contribution I can make is to approach the matter from the other end. If we know that we wish to assign valuations to something, whatever it is, and we know nothing else about it, then it doesn't matter if it is called an object, or an Ur, or a spin-operator. If the possible range of valuations that we are going to assign to it are, for example, exactly the possible range of valuations that we would assign to it in a classical logic then it could very well be called an object in quite an ordinary sense. But if they are essentially different as I believe them to be (and I have a great difficulty with some of the logical consequences, but never mind that for the moment) then I have no word whatever to use, because I am precluded from using a Machian type word such as 'observation', or even phenomenon— for these suggest that there is some kind of sensation or *Urexperienz* behind them, and there isn't. I now suggest something: namely that what makes it difficult to assign properties to these objects is because time is essentially involved in their existence. This is because the process of observation or measurement, even of a single variable, is essentially limited by the finiteness of signal velocities: it is essentially limited because it must take some time, and my own opinion is—without trying to go into detail—that this has most serious effects on the logic. Indeed I think these effects are much more serious than von Weizsäcker has suggested in his minimal modification of classical mechanics. It is the kind of modification of logic that was introduced for example by Brouwer, for similar reasons (and the paradoxes involved in observation are closely connected with mathematical paradoxes which in a certain sense are caused by mathematicians trying to observe their own processes). So there will be most serious changes in the logic, and these seem to me to be the as yet open questions of epistemology.

(VON WEIZSÄCKER): I agree. I think 'quantum logic' properly understood, is more radical than Brouwer's view on logic.

INDEX OF PERSONS

The Index of persons does not contain the names of authors of chapters in this book, except where they are mentioned outside their own contributions. *Italicised page numbers* refer to the bibliographical lists at the ends of chapters.

INDEX OF SUBJECTS